NATURAL BEHAVIOR

NATURAL BEHAVIOR

The Evolution of Behavior in Humans and Animals Using Comparative Psychology and Behavioral Biology

Burton A. Weiss, Ph.D.

Professor Emeritus, Drexel University
Adjunct Professor Emeritus, University of the Arts

Universal-Publishers
Irvine • Boca Raton

Natural Behavior: The Evolution of Behavior in Humans and Animals
Using Comparative Psychology and Behavioral Biology

Universal Publishers, Inc.
Irvine • Boca Raton
USA • 2018
www.universal-publishers.com

ISBN 978-1-62734-242-1 (pbk.)
ISBN 978-1-61233-141-6 (ebk.)

Typeset by Medlar Publishing Solutions Pvt Ltd, India
Cover design by Ivan Popov

Publisher's Cataloging-in-Publication Data

Names: Weiss, Burton A., author.
Title: Natural behavior : the evolution of behavior in humans and animals using
    comparative psychology and behavioral biology / Burton A. Weiss.
Description: Irvine, CA : Universal Publishers, 2018. | Includes bibliographical references
    and index.
Identifiers: LCCN 2017962734 | ISBN 978-1-62734-242-1 (pbk.) | ISBN 978-1-61233-141-6
    (ebook)
Subjects: LCSH: Human behavior. | Human evolution. | Animal behavior--Evolution. |
    Hominids--Evolution. | Psychology, Comparative. | BISAC: PSYCHOLOGY / Animal &
    Comparative Psychology. | SCIENCE / Life Sciences / Evolution.
Classification: LCC BF671 .W45 2018 (print) | LCC BF671 (ebook) | DDC 156--dc23.

This book is dedicated to

E. G. Wever

and

T. C. Schneirla,

two mentors and giants of comparative study.

Contents

Preface

The intent of this book is to present a concise history of scientific thought and the topic of evolution along with the understanding of the evolution of behavior from single cell organisms to humans with many original ideas. The ideas come from many decades of university teaching, research, and study.

Any attempt to cover such a large span of life must be a selective survey of example phyla and species because the field is immense. Any one species could occupy an entire career of study. Hopefully, this book will interest readers to pursue the topic of the evolution of behavior and specific species more intensively.

In acknowledgement, I want to thank my wife, Ruth A. Weiss, for her encouragement and enablement in the creation of this book. I also want to thank William J. Cook, Delaney Johnson, and Joseph Selm, for their assistance and advice.

To conclude, I am offering some of my animal poetry.

Rabbits
Rabbits have habits
of hopping around
my well tended ground,
while mirthfully devouring
everything flowering.

Ants
Ants have pants
that no one has seen
because they are not keen
to wrestle with the glitches
of fitting six legs into britches.

CHAPTER 1
The Subject of Nature

The Trend Away from the Egocentric View

Throughout the history of human thought, especially in Western civilization, there has been a clear trend. That trend has been away from human egocentrism, the tendency to view humanity's position in the world as primary. The trend has been led by science.

A review of the highlights of the trend could begin with the ancient Greek, Aristotle (384–322 B.C.E.), who organized the Universe with the Earth at the center and the celestial bodies revolving around it. The terrestrial world was divided into ranks ordered by degrees of perfection. The ranks placed people superior to animals and animals above plants. Each level had specific characteristics. Thus, people were rational while animals were instinctive.

About 250 B.C.E. Aristarchos of Samos proposed the heliocentric concept of the Universe with the Earth revolving around the Sun. Although, the heliocentric view was adopted by some astronomers like Seleukos of Seleukia, Aristotle's geocentric version prevailed and was preeminent into ancient Rome. After the end of the Roman Empire, Aristotle's work was lost to European knowledge. However, the civilizations of North Africa retained Aristotle's ideas, which were especially promulgated by the astronomer, Ptolemy of Alexandria, in the second century C.E. Ptolemy developed a geometric model of Aristotle's view of the central Earth and the Sun, Moon, Planets, and Stars revolving about the Earth in circular orbits. The model was cumbersome, but allowed calculation of planetary positions. Maimonides (Moses ben Maimon,

1135–1204 C.E.) incorporated Aristotelian ideas into Hebrew theology and from there they spread into Islamic doctrine. In the middle 1200's C.E. Ibn Daud translated Hebrew, Greek, and Arabic science, philosophy, and theology into Latin. The Latin translations returned the geocentric ideas of Aristotle to European civilizations, which were emerging from the dark ages. Aquinas (1225–1274 C.E.) read Maimonides and brought the thinking of Aristotle into Christian theology. He converted the linear levels into a system topped by absolute perfection, God. The next level was occupied by beings still too perfect to commit sin, angels. The following level held rational organisms, people. In the fourth level were instinctive creatures, animals. Finally, at the bottom of the order is vegetative life, plants. Modern Hebrew, Islamic, and Christian theology retain the image of rational people separate from instinctive animals.

However, Aristotelian geocentric views were questioned when Hasdai Crescas (1340–1410 C.E.) employed logic to refute them. Copernicus (Nikolai Kopernik, 1473–1543 C.E.) postulated and Galileo (1564–1642 C.E.) proved that the Earth was not the center of the universe or even the solar system, but actually orbited the sun. These discoveries, now elemental, were vehemently attacked. They were a challenge to the egocentric view of humanity as the center of the universe that had been incorporated into theology. Copernicus, fearful of persecution, hid his papers for posthumous publication. Bruno (1548–1600 C.E.) studied Copernican principles and unwisely espoused these ideas in public, for which act he was burned. Galileo was tried in Italy for heresy, forced to recant, and incarcerated for life in 1633 C.E. He was forbidden from writing further and all of his works were burned. But, Galileo's ideas had already spread beyond Italy.

So strongly held was the egocentric position that even seemingly remote challenges were met with severe sanctions. Thus, Servetus (Miguel Serveto, 1511–1553 C.E.) was burned by John Calvin for describing blood circulation as being pumped by the heart. The prevailing view was that blood ebbed and flowed like the tides as Aristotle had stated. People of the era thought the heart was the seat of the soul and conceived of health and personality as based on the balance of body fluids. We still retain many heart references for emotions. Two centuries later, Benjamin Franklin was also widely condemned by clergy for investigating such "heavenly" phenomena as lightning. That reaction to scientific exploration influenced the separation of church and state clause in the American Constitution, which Franklin helped compose. The framers of the Constitution were determined to prevent clergy from destroying people for having new ideas. Yet, the myth of devout founding fathers is still promulgated despite contrary evidence. For example, Washington opened his successful attack on Trenton and subsequently Princeton on Christmas. Had Washington been devout and not launched his campaign on Christmas, he would not have had the element of surprise necessary to defeat the better armed and trained opposition. Thus, the Colonies would have remained British. Washington would not likely have been surprised by an attack on Sunday morning like that at Pearl Harbor.

Darwin, in explaining the origin of species, challenged the extremely critical egocentric view of humanity as a direct, divine creation. Attacks on those teaching evolution are rampant and frequent even today, over a century and a half after Darwin's initial publication. Darwin's ideas will be considered in later text.

Einstein's contribution to the trend away from the egocentric position was contained in his principles of relativity. Relativity challenged many concepts of absolutes in the universe. Some clergy still confuse the physical principle of relativity with the unrelated philosophical and theological position of relativism, which depicts morality as relative. Thus, opposition to the supposed concepts of relativity has grown because of lack of understanding.

Freud, in turn, demolished the egocentric view of human rationality. After Freud's pioneering techniques of psychoanalysis have been superseded, Freud's demonstration of the non-rational foundation of human behavior will persist as a milestone in the trend away from the egocentric view of people.

The Nature of Science

Since such a major trend in human thought has been led by science, it is cogent to question the nature of science. Also, needing questioning is how the move away from the egocentric view could be pushed against very strong resistance to new ideas. Frequently, science is linked with a lengthy history of inquiry. Certainly, the older sciences like physics and astronomy have such a history of inquiry. However, many younger sciences barely have any history of inquiry, and a new science starting tomorrow would have none. In addition, disciplines clearly not sciences, such as art and music, have a history of inquiry into their own aesthetic analysis of the worlds with which they deal. An artist or musician usually must master the history of the subject before generating original work. Therefore, a history of inquiry is not what distinguishes a science from other types of endeavor.

Sciences typically acquire a body of facts and laws. The presence of a body of facts and laws is, therefore, sometimes used to identify a science. The same objection to this differentiation of science can be raised as was for the history of inquiry definition. Namely, new or young sciences do not have bodies of facts and laws, and non-scientific disciplines such as art and music do have bodies of facts and laws. An artist or musician typically masters the facts and laws of the subject before creating original work. In a portrait, for example, the eyes are half way down the head. Further analysis of the apparently cohesive body of facts and laws of even the older sciences shows some basic flaws.

Light, for example, has been a subject of study for a long time in the science of physics. Light is defined in the psychological, not physical, terms of electromagnetic energy stimulating the human retina resulting in vision. Related energy, like infrared or ultra-violet, are not visible to humans and do not count as light. Light is also treated on at least three levels, each with distinct mathematical analyses with separate understanding. Physicists deal with light as a ray or beam with geometrical optics for the topics of reflection, refraction, and similar phenomena. Light is also considered as a wave with sinusoidal analysis for interference, diffraction, and related occurrences. Finally, light is viewed as a packet, or photon, with statistical quantum analysis for phenomena like absorption and emission of light energy and comparable events. The cohesion of this body of facts and laws is only apparent. Thus, the attempt to differentiate a science from other disciplines by any accumulated body of facts and laws must be discarded and, the question of the nature of science remains.

Sciences always deal with specific observations of events in the world and with general hypotheses about the observations. But, so also, do many non-scientific disciplines like art and music. Scientists, however, have very definite ways of developing hypotheses from observations by the method of inductive logic and, in turn, of checking hypotheses against observations by the method of deductive logic. These methods of inducing hypotheses from observations of events, and deducing observations that should follow from hypotheses are the signatures of all sciences. Science is the scientific method. Knowledge is generated by the use of the scientific method. Figure 1-1 graphically depicts the scientific method of progressively cycling from observations to hypotheses to observations, etc. The cycling may be started anywhere depending on the particular subject and the abilities of the scientists involved.

Some sciences emphasize the deductive side of the cycle. Physics articles are generally published in the deductive form of stating a particular hypothesis, then describing an experiment which generated observations bearing on the accuracy of the hypothesis. Other sciences emphasize the inductive portion of the cycle. Astronomy depends heavily on careful observations to generate hypotheses about the stellar universe. However, all sciences employ the cyclic method, and any instance of the method being employed is science.

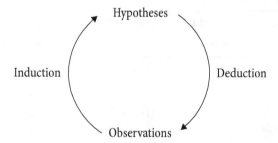

Figure 1-1 The Scientific Method of generating knowledge by employing inductive logic to generate hypotheses from observations and deductive logic to check the accuracy of hypotheses against observations.

Underlying the use of the scientific method is one basic assumption, scientific determinism. Scientific determinism is the assumption that events in the universe are predisposed toward lawful behavior. Without the assumption, science is reduced to dealing with unique occurrences unrelated in time or space. What is discovered in one place is not necessarily true in another location.

What is valid now may not have been yesterday and is uncertain tomorrow. With the assumption of scientific determinism, science studies continuing processes of events in the universe.

Scientific determinism is not the same as determinism, the philosophical idea of predestination or fatalism. The assumption of scientific determinism, that events in the universe are lawful, does not require such restrictive conditions. The only stipulation is that, all of many possible outcomes of an event are each in themselves lawful, even if at times unpredictable. Examples of lawful but unpredictable events are rolling dice, and the radioactive decay of atoms.

Dice tossing, tabulated in Table 1-1, represents the principle of scientific determinism, because all the possibilities are lawful, but not predestined. Thus, a score of seven has six possible ways to occur (five plus two, two plus five, three plus four, four plus three, six plus one, and one plus six) out of 36. The probability of a seven occurring is six in 36 or once in every six tosses. In a population of 36,000 tosses there will be 6000 sevens with very little error, and much less than a percent variation. However, if on any single toss, anyone wishes to know whether a seven will appear, only the estimate of probability (a one in six chance) can be invoked to predict. There is no certainty

Table 1-1 Tabulation of Dice Tossing.

Score	Possible Ways	Probability	Frequency (N = 36K)
2	1	$1/36 = .0278$	1000
3	2	$1/18 = .0556$	2000
4	3	$1/12 = .0833$	3000
5	4	$1/9 = .1111$	4000
6	5	$5/36 = .1389$	5000
7	6	$1/6 = .1666$	6000
8	5	$5/36 = .1389$	5000
9	4	$1/9 = .1111$	4000
10	3	$1/12 = .0833$	3000
11	2	$1/18 = .0556$	2000
12	1	$1/36 = .0278$	1000
Totals	36	$36/36 = 1.0000$	36,000

of predestination in individual tosses, only an estimate of probability. But, again, whatever score occurs will be a lawful event consistent with scientific determinism.

Natural phenomena, like the decay of atoms, follow the same principles as dice tossing. In a population of atoms, like a population of dice tosses, decay will follow a predictable course. Whether a particular individual atom will decay at a given moment, however, is a matter of probability, just like a single dice toss. Einstein's comment that, in atomic phenomena, "God does not play dice" ("Gott würfelt nicht") was premature, but the atomic dice are used in a subtle manner, as Einstein also observed "God is subtle, but not malicious" ("Raffiniert ist der Herrgott, aber boshaft ist er nicht."). Tracing of the understanding of the principles of evolution, later in this chapter, also reveals the role of scientific determinism.

The Science of Nature

Because the scientific method is shared by all sciences, science is essentially indivisible. Any separation among sciences is arbitrary and splits fields of study. The common demarcation between social sciences and biological sciences splits psychology. On the other hand, segregation of sciences along lines of life sciences and physical sciences divides biochemistry. Separation of sciences, however, is frequently necessary for convenience in dealing with the wide array of subjects under the scrutiny of science. This book, concerned with nature would, of course, deal particularly with those sciences studying life and not especially with those sciences whose subject is the physical world.

A basic difference between life and the physical world is the importance of the time dimension. In the physical universe events in time have relatively long duration compared to the short span of life. In view of the small importance of time in the physical world, physics has not investigated time to any great extent (Gold, 1967), until the recent spate of work over the last thirty years. Life, however, existing as it does in a rotating world, which makes the main energy source, the sun, periodic, has become time-locked. Even basic metabolic processes exhibit time cycles. Time brings rapid and important changes to life and can be considered life's most salient dimension.

Physics considers work equivalent to force acting through distance (W = Fd), while force acting through time is relegated to the status of impulse (I = Ft). Such definitions are confusing to students of life science who realize that tremendous energy expenditure can result from holding a weight in fixed position for a time (impulse) as well as from lifting a weight through a distance (work). Actually, a living organism performs work by changing chemical potential energy into kinetic energy, just to maintain its existence from moment to moment. Indeed, the best index of energy expenditure during motor behavior is the integral of force acting through time (Trotter, 1956). Force exerted through time is also a major parameter of an organism's response repertoire (Notterman and Mintz, 1965). Thus, for the purpose of life sciences, work must be reconsidered as effort acting through both distance and time, Effort = F[(d/t) + t]. The range of possible values of space (d) and time (t) are limited by the capacity of the organism. Within that range there are optimum values for d and t. To vary slightly from the optimum performance greatly increases the effort. Power (Fd/t) plus impulse (Ft) are the expression of effort for organisms. Beginning physics students have difficulty with the physical concept of work because of the intuitive understanding of their own effort. Technology has increased human power by decreasing the time of various activities. Using tools also changes the sense of how much effort is required for a task. The formulation of effort is realistic for the life sciences because both time and space become important to an organism that moves through its life.

A popular view of the difference between life sciences and physical sciences is that life sciences rely on statistics and physical sciences employ laws. The view stems from the idea that life sciences are newer, not as established, and understand less of their subject than the physical sciences. Implied in these statements is the concept that statistics is only a temporary treatment awaiting more thorough knowledge of the subject and the subsequent formulation of laws.

While the statements have some degree of accuracy, the stopgap view of statistics is incomplete. Physical sciences do use statistics. Returning to the example of the phenomena of light, laws are employed when light is treated as a ray. Thus, angle of incidence equals angle of reflection. Laws are evidenced when light is viewed as a wave. Hence, Huygen's principle stating that every point on an advancing wave front is a secondary wave source. However, statistics must be be invoked when light is considered as a photon. Thus, events, like a particular photon being absorbed by colliding with a specific electron at just the proper moment, although lawful, are necessarily only probable. Probabilistic events can only be treated with statistical estimates of the chance of their occurrence.

The example becomes even more significant by illustrating the essential difference between laws and statistical analysis. Light rays or waves are populations of photons. Laws always describe the behavior of populations. Thus, laws can describe the motion of a ball down an incline because the ball is a population of molecules. Laws can predict that only a fraction of the population of depositors of a bank will want their money at once. That enables the bank to retain just a small amount of reserve capital and invest the rest. Laws can predict the frequency of scores in the 36,000 dice tosses in Table 1-1. Whenever an individual in a population is singled out, however, the event becomes probabilistic, and a statistical estimate must be used. Individual molecules of the ball rolling down the incline could scrape off on the surface or, even, sublimate into the air. Individual bank customers could request their money. Single dice tosses are unsure. Formulation of laws or use of statistical analysis depends not on the science, but on whether the event in question is a population phenomenon or involves an individual. For example, to measure the individual size of leaves on a tree a statistical sample is employed, but to compare the populations of oak and maple leaves, only a few are needed.

The province of the life sciences is life. But, life with its conservatively estimated 1,200,000 animal species (Hanson, 1964) and 333,000 plant species (Arms and Camp, 1989) is still a relatively small population compared to phenomena in the physical universe. The ball that rolled down the incline contains more molecules than the total population of all of life's individuals on this Earth. Thus, life scientists investigating small populations, and often individuals, must frequently employ statistical techniques.

In addition to small populations, life sciences must also explain the high degree of complexity and organization of life. Physical parameters like size do not help. Measuring species size does not cope with the differences among 100 foot whales, 300 foot tall sequoias, and three-micron wide microsporidian spores. To understand the complexity and organization of life, life scientists must know the history of life. They need to comprehend life's origins and evolution to its present state, or how life came to be the way it is. Evolution provides the historical analysis and deterministic mechanism, in natural selection, that unveil the phenomena of life. Thus, evolution is the fundamental underlying principle of all the life sciences.

History of Evolutionary Theory

The concept of evolution involves the change of species through time. A comprehensive review of the history of evolutionary theory would fill volumes. However, a synopsis, of some of the major steps in the history of evolutionary theory, serves to illustrate the creative and dynamic effort of many great thinkers in formulating modern evolutionary theory, and to clarify past errors.

Among the earliest theories proposing change of species was Aristotle's linear hierarchy, already noted on the first page of the chapter. In this system each level was more perfect than the level below. Thus, in the version of Aquinas, God represented absolute perfection followed by angels, who were less perfect than God, but still too perfect to commit sin. The next rank, at the top of the real world, was humans. Humans were less perfect than angels, in that people were rational and could choose to commit sin. Heaven was reserved only for those who did not sin. After humans were animals. Animals were viewed as even less perfect because they could not reason, but only react with instinct. Finally, came plants, which were at the bottom in perfection because they merely vegetated. In addition, those in each level strove to be more like the level above. The level of understanding of life for most people, even today, reflects the idea that humans are rational, but animals respond only by instinct. The Hindu concept of the scheme of life inherent in reincarnation, also includes a similar ordering of life. It contains the idea of transmigration that includes all life in a pursuit of perfection and heaven, or nirvana.

Galileo developed a theory of impetus which held that objects maintain momentum unless stopped. The theory was influential because it avoided causal or animistic explanations for movements. Momentum kept the planets in orbit, not some unknown invisible gear system that had to be cranked. Change, not stability, was the natural state of the universe. Coupled with other later discoveries, like calculus by Leibnitz and Newton, Galileo's ideas triggered a 17th century revolution in thought, which emphasized mechanism and scientific determinism. The revolution led to great progress in the physical sciences and caused abandonment of the teleological view that the complex organization and design of the world called for the existence of a designer. Darwin was later to forsake the same teleological, no-clock-without-a-clock-maker, approach in his explanation of the origin of species. However, the teleological idea is still influential in theology.

Life science, in awe of the progress of the physical sciences, began to emulate them. Led by Linnaeus (Karl von Linne) in the 18th century, life scientists developed a complex classification system, categorizing species by anatomical and functional similarities. The rationale was that classifying would reveal organization and permit induction of laws concerning life. But, species were considered immutable entities.

Hegel originated the technique of using history as an analytic tool, rather than merely as a description of past events. He was interested in the origin of social institutions and structures. His thought had tremendous impact on the social disciplines of history, anthropology, economics, sociology, and social psychology. Hegel's new analytic approach also began a 19th century revolution in other sciences. Astronomy and physics began using Hegel's approach to explain the origin of the universe and to conceive astrophysics. Lyell employed the analytic approach in geology by using the history contained in rocks to explain the formation of the Earth. His work heretically replaced the story of Noah and the flood to explain the origin of the Earth's structures. Lyell's thought influenced the closely related field of paleontology and a young student at the time, Charles Darwin.

New ideas were accumulating rapidly in the life sciences. Buffon, in 1760, had destroyed the purposive explanation of organs, like eyes are for seeing. He observed that two of the five digits on each of a pig's feet did not touch the ground. Therefore, toes could not be for walking. The pig's vestigial digits led Buffon to explain all quadrupeds as degradation from fourteen primary types. Animals, instead of being direct divine creations, were now viewed as imperfect descendants of originally divine creations. The theory cracked the ground of life science for later seeds of historical analysis of life's origins.

By 1803 Lamarck, Buffon's student, had formulated the first theory attempting to explain life as progressive and adaptive change. Life diverged from a primal type, rather than degraded from divine forms. Lamarck's theory held that acquired characteristics, like large muscles, affected the (at that time unknown) genetic material and were inherited. Thus, a right-handed blacksmith's son should have a muscular right arm. Many a blacksmith's son did have a large arm, but not because of his father's acquired characteristics. Only in rare instances, like alteration of the sex cells by drugs or radiation, can acquired changes be inherited. Lamarckian theory could not explain how a blacksmith, who lost an arm, could subsequently have offspring with complete arms. Lamarck also used the old Aristotelian notion of striving toward perfection. He contended that the striving toward perfection of organisms led to the acquisition, through use, of improved characteristics, that were, in turn, inherited. Disuse would result in the loss of characteristics and the subsequent reduction of those characteristics in future progeny. Although such concepts mediated Lamarck's impact among life scientists, his influence was still immense. Indeed, there are even recent Lamarckians, like Lysenko (Morton, 1951), the former President of the Lenin Academy of Agricultural Sciences. Lamarckian theory adapted readily to the milieu of socialist realism in which Lysenko worked. Against the advice of knowledgeable experts like N. I. Vavilov, who was later purged and died in detention, over 3000 biologists, who disagreed with Lysenko, were dismissed, imprisoned, or executed. Lysenko made decisions affecting grain breeding in Russia. Lysenko's influence, destroyed progress in Soviet genetics and crop production. Russian influenced countries like Czechoslovakia, Poland, China and others were also affected by Lysenkoism. After Stalin's death Lysenko lost his position, and Soviet genetics again flourished. Khrushchev reinstated Lysenko, but on February 4, 1965, Lysenko followed Khrushchev in being ousted. Lysenko died in Kiev on November 20, 1976.

Lysenko illustrates the periodic resurgence of Lamarckian ideas. Historically, however, Lamarck's theory experienced difficulty from the beginning. St. Hilare, Lamarck's student, debated Cuvier in the French Academy of Science in 1830. Cuvier favored the idea that the world reached its current condition as the result of a series of catastrophic changes. Each change gave major reorganization to life forms. No mechanism for gradual accumulation of alterations was included. St. Hilare supported Lamarck's position, but used evidence linking the independent and unrelated, though similar, eyes of vertebrates and cephalopods, like the octopus. Cuvier, being knowledgeable in anatomy, caught St. Hilare's error and won the contest. The debate was extremely influential and served to suppress ideas on evolution for the next three decades. At the time, the poet Goethe was reportedly more interested in reading the result of the debate than in seeing a critical review of his own poetry. Darwin was a college student at the time. In addition to the debate and the work of Lyell, Darwin also pondered the ideas of Malthus. Malthus contended that the rapid reproduction and subsequent increase in the population of organisms would outstrip food supply, unless there were checks on the population. The next year, 1831, following graduation, Darwin joined the H. M. S. Beagle as ship's naturalist for a five-year voyage around the world to survey, record, and collect plant and animal specimens.

In 1859 Charles Darwin published his book, *The Origin of Species*. The book was the culmination of his observations on the Beagle journey and of his subsequent thought. He had presented his ideas the previous year before the Linnaean society in London along with Wallace, who offered similar concepts. The competition with Wallace finally led Darwin to publish. His work was so well anticipated that it sold out on the very first day.

Darwin had enormous impact on life science because he presented not only the idea that the evolution of life had occurred, but he also explained the process with the mechanism of natural selection. In addition, his wide travel experience enabled him to support his theory with an unequaled and unassailable array of accurate examples. The timing also favored Darwin's ideas about evolution. In 1860, Pasteur demonstrated conclusively that organisms were responsible for spoilage and fermentation, and that spontaneous generation of life did not occur. Virchow reformulated the earlier postulates of Schleiden and Schwann, that cells were the basic unit of life, and that cells came from other cells. The cell theory, coupled with Pasteur's discoveries and Darwin's work, sparked a major revolution in life science. For the first time, both the units of life, the cells, and the explanation of the origin and continuity of life were known.

But the heralds were not trumpeting. Pasteur was rejected. Not until later in the century, when called upon by the Russian Czar to stop an anthrax epidemic, would Pasteur's concepts be accepted. The cell theory was ignored until 1895, when Verworn revitalized it. Darwin's ideas became distorted into cliches like "struggle for existence" and "survival of the fittest." These phrases were wrongly employed to emphasize the survival of individuals, rather than the population principle of differential reproduction of the species. Spencer, and other Social Darwinists, usurped these cliches to excuse the inequities of the very rich and very poor of the rising industrial society. Thus, the wealthy were seen as the fittest and the poor were viewed as the unfit. Life scientists, horrified by the political distortion of their work, abandoned the pursuit of general explanations of life, like evolution. Again, laws describe the behavior of populations, not individuals.

For the next six decades evolution was taboo and reductionism reigned. Müller's followers tried to use his laws of nerve function to explain all behavior. Loeb joined the movement with his idea of taxis, the approach or avoidance of physical stimuli. He attempted to explain all behavior as the physics of taxis. Thus, a cockroach ran into darkness because of negative photo taxis. A resurgence of Lamarckian ideas, which focused on the individual, had prevailed.

Part of the difficulty was the fact that Darwin's work was lacking an important piece necessary to complete the puzzle. Darwin understood natural selection. He did not grasp the significance of sex as the mechanism for generating and preserving the population differences on which selection operates. Darwin believed that sex was a means of obscuring population variance. He thought offspring would be intermediate compared to parental extremes. Thus, a tall and a short parent should have a medium-size child. Darwin's ideas on natural selection were remarkable considering the paucity of genetic knowledge at the time.

In the 20th century, the, link that converted evolutionary theory to fact appeared. Mendel in 1866 had found the basic laws of genetics stating that parental characteristics do not blend in progeny. He found that the combination of various parental characteristics in the offspring followed the laws of statistical probability, and that dominant parental characteristics mask recessive ones in the next generation. Mendel's work, published in the Proceedings of the Brünn Natural Science Society (1866), was ignored until rediscovered independently in 1900 by Correns, deVries, and von Tschermak-Seysenegg. In 1903, Sutton realized that the chromosomes life scientists had been watching divide and reproduce through microscopes for years were the carriers of the genetic characteristics. In 1909 Nilsson-Ehle extended the understanding of qualitative gene characters, like red or white, to quantitative features having multiple genes, like the degree of redness. Chetverikov, Haldane, Fisher, and Wright in the late 1920's and early 1930's provided the mathematical basis for selection. Fisher (1930) synthesized evolution and genetics by indicating that genetic fitness was related to population variance.

Finally, life scientists knew the three features by which sex contributed to evolution. First, sexuality was a device for generating variability by rapidly creating various, and even new, parental genetic combinations in offspring. Sexual combinations were fast, compared to the slow accumulation of new features in asexual populations. Variation is also enhanced by dominant genes shielding, retaining, and preserving the variety of recessive characteristics, even, if they were fatal by themselves. Variation is important because it provides the resilience in the population to adjust to changes in selection. Second, generation of new genes by mutation, discovered by deVries, is a random occurrence with no regard for any needs of the organism, unlike Lamarck's purposive acquired characteristics. Third, the unit of evolutionary change is not the individual, as Lamarck thought, but the population's total pool of genes. To repeat, as with dice tosses, laws describe the behavior of populations, not of individuals. Sex itself is an illustration of the last point because sexuality, while important to individuals, has no adaptive significance to the individual. Males are no more or less adaptive than females. If they were, that would mean selection could operate separately on males or on females and eliminate one sex. That is not possible! The species, as a whole, is the unit of selection. Sex is a mechanism that has been favored because of its adaptive value of contributing to variation in the population, the real unit upon which selection operates.

Sex has proved so adaptive that the mechanism has appeared in numerous species in different individual forms. The male differentiating system (XY) in humans, the female differentiating system (WZ) in birds, and the haploid male (developing by parthenogenesis) and diploid female bees are all different. Thus, genetically, a human male is more like an avian female. However, mathematically, the sexual systems serve the same purpose of generating variation for the species. All sexual systems function alike to provide a source of variation in the population.

Some theories about the origin of sex suggest that early organisms were asexual like modern protists. With a functional method of reproduction already in existence, sex would have needed some purpose other than reproduction to displace the asexual technique. Indeed, enhancing variation in a population was just such an advantage. Sexuality became favored in many organisms, because it served to accelerate evolutionary rates by increasing variation. However, in some species sex promoted too much genetic variation. Thus, to slow down their evolutionary rate and adjust to the environment, those species abandoned sex by becoming hermaphroditic.

An alternative sequence could have been that the sexual mechanism was originally an exchange of genetic material between equivalent organisms like the conjugation of similar individual Paramecium. Specialization of sexual organs in multicellular species led to hermaphroditism. Interchange of genetic material was still between equivalent individuals employing self fertilization to adjust evolutionary rates, like many species of worms today. Sexual specialization of individuals into separate males and females then occurred through secondary loss of the organs of the opposite sex by hermaphrodites. Genetic interchange can be accomplished effectively by either hermaphrodites or separate sexes.

Separate sexes trade away redundancy to gain efficient reproductive specialization in individuals, but do not fundamentally alter the sexual mechanism. In this view, abandonment of the sexual mechanism has not happened, except partially, in some species like those in worm phyla which have returned to or retained asexual fission. Asexual fission by-passes the sexual functions of hermaphroditic reproduction and thereby adjusts the rate of genetic exchange.

In either theory, sexual forms were generated by having one switch with two positions. Thus, in humans, the switch is the hormone testosterone. Testosterone presence in early development means male, while testosterone absence means female. Humans are not basically female with males needing an additional hormone, rather one hormone switch simply determines development. Two switches, one for each sex, are not required. Natural selection has a tendency to be parsimonious.

Modern Evolution

Modern life science recognizes the fact that evolution depends on genetic, phenotypic (constitutional), and ecological opportunity. But the combination of opportunities in the right place and time is dependent upon probabilistic chance. Therefore, evolution is a matter of probability. The probabilistic nature of evolution makes it a one-way sequence and precludes an exact repetition of the process. If everything could be restarted, the same opportunities would not be likely to recur, especially in the same sequence.

Operating on the products of the combination of opportunities is natural selection. Natural selection determines the viability of any species by selecting which combination of genes will contribute most to the next generation. Thus, selection causes a drift in the proportion of gene alleles in a population toward dominance or loss. Even a very small advantage will result in major changes. Pauling (1968) calculated that a mutation rate as small as one in 20,000 per gene generation coupled with a minute advantage like 0.01% more progeny for the mutant would result in replacement of the standard by the mutant within one million years, for many primate species.

The conditions for evolution are defined by the Hardy-Weinberg law, summarized next in Table 1-2. The law states the circumstances which, if met, would preclude evolution. Any isolated population meeting the requirements of the law, without immigration or emigration, would not change because the effects of natural selection are negated.

Table 1-2 Summary of the Hardy-Weinberg Law.

Evolution will not occur in an isolated population with no immigration or emigration if:
1. There is an infinite population,
2. There is no or an equilibrium of mutation,
3. There is random reproduction, which requires,
   Equal survival
   Equal fecundity
   Random mating

The first condition required for no evolution is an infinite population. Since any change is finite, an infinite would be unaffected by any amount of finite change. Clearly, there are no infinite populations, but the first condition also indicates large populations would be more resistant to change. Thus, a large population could be threatened with extinction by not being able to change quickly and could be as endangered as a group of rare organisms.

The second postulate needed to prevent evolution is no mutation or the equivalent, an equilibrium of mutation. In an equilibrium, any mutation in one direction is balanced by back mutations by other genes. Without mutation, no original genetic material can influence a population. However, no population is without mutations and an enormous amount of mutation has already occurred in all populations.

The third requirement that would prevent evolution is random reproduction. Random reproduction requires three elements. One is equal survival of individuals through their reproductive period of life. Otherwise, unequal survival would produce differential reproduction. Unequal reproduction is how survival effects evolution, not individual struggle. A second element is equal fecundity of all individuals, because unequal ability to reproduce in the population would not be random. Finally, completely random mating is necessary for random reproduction. Nonrandom mating introduces a direction to reproduction.

Any living population necessarily violates these conditions and is, therefore, under selection pressure to change, resulting in evolution. The more a population deviates from the conditions of the Hardy-Weinberg law, the faster that population is evolving.

Survival is a consideration only in the final requirement of the law and then strictly as survival to reproduce, not as individual survival per se, like the "survival-of-the-fittest" idea. Just as in dice tossing, laws apply to the population not the individual. Evolution is differential reproduction of the population, not individual survival. The most fit to survive individual, without reproduction, has no direct influence on evolution.

The fact that evolution is population drift towards the composition of the reproducers is not an argument for uncontrolled, or even any, mating or for large families. Some organisms, like many species of fishes, do rely on sheer fecundity, by leaving thousands of fertilized eggs behind, to insure some adults for the next generation. Natural factors like temperature change and predation keep those populations in check. Other species, however, have evolved a parental care system, in which, one or both parents (or substitute adults) remain with the young, protecting them until adulthood. A parental care mechanism requires small families to avoid dissipating protection by covering too many young. Childhood accidents remain a major danger of elimination in human families, indicating the need for the parental care mechanism. In addition, an individual may indirectly influence the success of the species by contributing to the accumulated cultural heritage, whether or not, the individual participates directly in determining the next generation through reproducing. Many species, like ants and humans, are heavily dependent on non-reproducing individuals.

An understanding of evolution as differential reproduction permits consideration of some prevalent ideas. Viewing the distribution of people throughout the world, the fact becomes apparent that the vast majority of humanity is lower class. Thus, the lower class is the main source of people for the next generations. With an expanding population, each generation is composed of more and more lower class. From these facts the conclusion can be drawn that human beings, as a species, are undergoing diminished adaptive abilities. There are two aspects to such a conclusion, the qualitative and the quantitative. Qualitatively, evolutionary principles can not justify the idea of less adaptive abilities in the lower class, because class distinctions are based on economic, social, and political values, not on genetics. Genetically, there can be no difference which class is the main producer of the next generation, unless class distinctions are genetically based. Quantitatively, however, the conclusion has merit because the lower class does not limit family size. The result of lower class reproduction has been a booming human population. The increasing numbers invoke the concept of the first condition of the Hardy-Weinberg law and illustrate how large populations can become endangered.

Another common idea is that there is a genetic basis for the incest taboo in human societies. However, the incest taboo is not defined genetically. Some societies allow uncle-niece but not aunt-nephew marriage. Genetically, these marriages are the same. The definition of incest also varies. Some states permit first cousin marriages, but others do not. Incest taboos existed long before humans had any significant knowledge of genetics. The taboos really reflect the role of marriage in society. Marriage was an alliance of families. If a son and a daughter in the same family married each other, that would prevent two alliances with other families and weaken their own family. Only if the family were already powerful, like royalty, could intra-marriage be useful in retaining, rather than dissipating, the power of the family. Thus, the rules for royalty were to marry their relatives, namely, other royalty.

The practice of intra-marriage among royalty is taken as proof of the genetic basis for the incest taboo, because the Royal Family of England, and its relatives, experienced repeated occurrences of hemophilia in its lineage. However, if a family has no genes for hemophilia, incest will not produce it. The consequences of incest depend on the genes of the family. Positive results of incest are responsible for the development of crops and breeding of animals, like dogs and horses.

Incest is assumed to be genetically detrimental, because mating with close relatives causes more pairing of recessive alleles than would a random match from the general population. However, human mating is not random. We tend to pick mates who are like us and, therefore, are related or at least share many genetic characteristics. Each person has two parents, four grandparents, eight great grandparents, etc. The expansion of ancestors doubles for each past generation. However, in a few generations the number of possible ancestors (2^n) exceeds the population of the Earth in the past. Thus, we all must share common ancestors and be related. Most importantly, the idea that pairing recessives is bad comes from the idea that recessive is evil and dominant is good. Dominant and recessive are merely terms expressing which gene will evidence in the phenotype, when they occur in combination. Dominant and recessive do not mean good and evil, because the mutations that produced the genes were random and without regard for the needs of the organism. Natural selection determines the adaptive value of the alleles, not human social values. Positive, negative, or neutral consequences from incest would depend on the genetic lineage of the mates. A genetic pairing with detrimental recessives would tend to evidence those characteristics in incest. But a positive lineage would evidence those positive characteristics. Thus, the incest taboo has no genetic basis, but was produced by the important social values of human mating.

Significance of Evolution for Behavior

Hardy-Weinberg also explained the normal distribution of structural characteristics, like height and weight, in terms of the possible gene combinations at mating. Thus, for one pair of genes, the binomial expansion, $(A + a)^2$, gives the frequency of possibilities ($A^2 + 2Aa + a^2$). A and a are the probability of occurrence of each gene allele ($A + a = 1$). With increasing numbers (n) of gene alleles, and estimates of the numbers of gene alleles in a population can be large, the frequency of possibilities $(A + a)^n$ approaches the normal distribution. For an imaginary, because all organisms have thousands of genes, single-gene organism with one dominant (A) and one recessive (a) gene allele, the mating possibilities of the population can be seen in Table 1-3.

Table 1-3 Gene Frequencies for Aa Type Parents.

		Female Parent	
		A	a
Male Parent	A	AA	Aa
	a	Aa	aa

The tabulated totals for Table 1-3 are: 1AA + 2Aa + 1aa. Graphically represented, the tabulated totals would look like Figure 1-2.

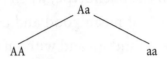

Figure 1-2 Graph of the Tabulated Frequencies of 1AA + 2Aa + 1aa.

For another imaginary, again, because all organisms have thousands of genes, dual-gene organism with two dominant (A, B) and two recessive (a, b) gene alleles, the mating possibilities of the population can be seen in Table 1-4.

Table 1-4 Gene Frequencies for AaBb Type Parents.

		Female Parent			
		AB	Ab	aB	ab
	AB	AABB	AABb	AaBB	AaBb
	Ab	AABb	AAbb	AaBb	Aabb
Male Parent	aB	AaBB	AaBb	aaBB	aaBb
	ab	AaBb	Aabb	aaBb	aabb

The tabulated totals for Table 1-4 are

1AABB + 1AAbb + 2AABb + 2AaBB + 4AaBb + 2Aabb
+ 2aaBb + 1aaBB + 1aabb.

Graphically represented, the tabulated totals would look like Figure 1-3.

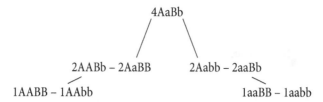

Figure 1-3 Graph of the Tabulated Frequencies of 1AABB + 1AAbb + 2AABb + 2AaBB + 4AaBb + 2Aabb + 2aaBb + 1aaBB + 1aabb.

With just the two genes graphed in Figure 1-3, the shape is beginning to arrange into the familiar "bell curve" of the normal distribution. Increasing the number of genes determining a characteristic to a great many, as is typical, produces a normal distribution. Modern knowledge of genes is that they also interact and turn on and off, resulting in complex function. Thus, gene content results in the normal distribution of characteristics.

Life sciences typically employ structure to document the evolution of species, because structure leaves bodily or fossil records. Genes determine structure, which, in turn, determines behavior. Human genes mean we have arms, not wings, and that determines our behavior. But what most people, even many life scientists, fail to realize is that behavior determines the genes of the next generation through reproduction. Natural selection is the environment, which has no way of interacting with genes or structure. What natural selection actually selects is neither the genes that determine structure nor the structure that determines behavior, but the behavior itself. The behavior of an organism is what interacts with the environment. The behavior of the organism interacting with the environment selects which organism reproduces and, thus, establishes the genes, subsequent structure, and consequent behavior for the next generation, as illustrated in Figure 1-4. The evolution of any species is, therefore, as much an evolution of behavior as of structure. Behavior is, actually, the central feature of evolution because it is the focus of selection.

Therefore, behavioral characteristics, like aptitudes, also display a normal distribution. Behavior, structure and genes are inseparable in evolution. The evolution of a species could be traced by its behavior as well as by its genes or its structure. But, following the evolution of an organism by its behavior, using

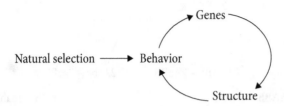

Figure 1-4 The cycle of the behavior of the organism interacting with the environment, selecting which organism reproduces and, thus, establishing the genes, subsequent structure, and consequent behavior for the next generation.

the fossil record is exceedingly difficult, because the behavior died with the organism. However, some behavior can be inferred from structure, like large canine teeth indicate a predator and extensive molars mean a grazing animal.

Courtship pattern is an excellent example of the evolution of behavior. Successful courtship leading to reproduction is extremely adaptive and highly favored by natural selection, because that behavior produces the next generation. Unsuccessful courtship, like mating across species, is selected against because such behavior creates, at most, sterile hybrids and does not contribute to the next generation. Any structure or behavior that leads to sexual attraction and subsequent successful courtship is favored. Repeated favoring of such advantages has led to very elaborate mate-attracting structures and extremely complicated courtship rituals in many species. The intricate courting "dances" of many birds are a dramatic example of the evolution of behavior.

Darwin (1871) was puzzled by mate choice because he thought that such behavior provided a direction in evolution separate from natural selection. Modern theorists also give a separate role to sexual selection (Daly and Wilson, 1978). Trivers (1972) felt that parental investment in offspring would influence sexual selection, even in humans. Thus, males would tend to be more polygamous, but females, with the heavier investment of pregnancy and infant care, would be more monogamous. Concluding that sexual selection is strictly genetic is impossible. Again, genes determine structure and structure influences behavior, which results in the genes of the subsequent generation. Mate choice is strongly influenced by behavior and the cultural consequences. In addition, choice of mates is part of the third condition of the previously discussed Hardy-Weinberg Law, in that, it violates the process of random reproduction. Without random reproduction, change in the population will occur and evolution accumulates through sexual selection.

Mate choice can also become out-of-phase with the environment, leading to conflict with natural selection. The Irish Elk (Megaloceros) is an example. Mate choice led to larger and larger male antlers, which had to be shed and regrown each year. The required energy expenditure produced extinction. Natural selection is the instrument of evolution.

Sexual selection, like any structural or behavioral feature, can be adaptive or not. Thus, sexual selection is not separate from evolution. Not only courtship patterns, but all behavior evolved. The process and phenomena of the evolution of behavior are the subjects for the remaining chapters.

The Non-conflict with Religion

One of the major reasons for difficulties in the trend away from the egocentric position noted early in this chapter is the supposed conflict with religion. If religion is ultimately the belief in God, rather than in human doctrine, there can be no conflict with science. Science can neither confirm nor disprove the existence of God. God is either believed in or not.

Difficulty stems from those who profess to believe, but whose personal faith requires proof, often from physical sources. The support for such fragile faith frequently comes under the scrutiny of science and the frail faith becomes threatened. The true believer knows that science, whose purview is phenomena, and religion, which deals with faith, do not even overlap and, therefore, cannot possibly conflict.

Clergy, however, repeatedly make ridiculous spectacles of faith by pronouncements about science. In 1650, Archbishop James Ussher of Armagh, Ireland, calculated from the ages of biblical figures that the world was created on Sunday, October 21, 4004 BCE, at 9:00 AM. Early in the 20th century, when several groups were struggling to develop "heavier-than-air" flying craft, clergy announced that such endeavors were folly because, if God meant people to fly, we would have had wings. The Wright Brothers on December 17, 1903, were the first to prove that we could make our own wings. Repeated interference in science by uninformed clergy have created an image of religion as anti-science. Instead, the misunderstanding is produced by the weak faith of the particular clerics. No one with strong faith is disturbed by evidence that the Earth is not the center of the universe, that humans evolved, or any other phenomena of the natural world.

If someone could be sent back a few centuries in time with the task of telling people of that time about nuclear energy, and saying that those people should write down what is explained to them, what would they write without any knowledge of what they are being told? Thus, anyone, thousands of years ago, being told how the universe was created would have no understanding of what to transcribe. Having no numbers like seven billion, they might write seven days and compose a description like Genesis. Therefore, the Bible describes what happened and science is trying to explain how.

References

Arms, K. and Camp, P. S. (1979). *Biology*, Holt, Rinehart and Winston, N. Y.

Daly, M. and Wilson, M. (1983). *Sex, Evolution, and Behavior*, PWS Publishers, Boston.

Darwin, C. (1887 edition). *The Descent of Man, and Selection in Relation to Sex*, D. Appleton and Co., N. Y.

Fisher, R. A. (1930). *The Genetical Theory of Natural Selection*, Claredon Press, Oxford.

Gold, T. (Ed.) (1967). *The Nature of Time*, Cornell University Press, Ithaca, N. Y.

Hanson, E. D. (1964). *Animal Diversity*, Prentice-Hall, Englewood Cliffs, N. J.

Morton, A. G. (1951). *Soviet Genetics*, Lawrence & Wishart, London.

Notterman, J. M. and Mintz, D. E. (1965). *Dynamics of Response*, John Wiley & Sons, N. Y.

Pauling, L. (1968). "Orthomolecular Psychiatry", *Science*, **160**, 265–271.

Trivers, R. L. (1972). "Parental investment and sexual selection", In B. Campbell (Ed.), *Sexual selection and the descent of man 1871–1971*, Aldine, Chicago.

Trotter, J. R. (1956). "The physical properties of bar pressing behavior and the problem of reactive inhibition", *Quarterly Journal of Experimental Psychology*, **8**, 97–106.

CHAPTER 2

Nature as the Subject

The Study of Nature

The study of nature is undertaken for many reasons. Humans have always had an interest in nature as the cave paintings by early people illustrate. People often identify with animals particularly in England, and with Disney encouragement, in the United States. Animal and human qualities are often intermixed as in mythology, astrology, and augury. People are also interested in learning how to use animals for food, clothing, and other products or as pets. However, the scientific study of nature is concerned with gaining knowledge as an aid to understanding organisms in their own right. Comparative study of many species, can lead to the discovery of general principles explaining the nature of all life. Practical considerations, like the application of bat sonar techniques to electronic guidance and detection systems, are the consequences, not the goals, of the basic study of nature and are possible only because of fundamental understanding.

Scientific understanding of nature comes from the use of the inductive and deductive methods in the pursuit of knowledge of nature. However, as in any human endeavor, scientists themselves often become embroiled in specific issues that retard progress. For nearly five decades, from about 1920 to 1960, a specific school of psychological thought called "learning theory" or behaviorism dominated the American psychological journals, excluding other work on animal behavior and wrongly emphasizing a strictly deductive approach employing laboratory studies and group statistics necessitated by the use of a single species, typically rats, but also pigeons and college students. When attempting to extract information in a study employing one species, statistics are required in order to discriminate between intra-species variation in the subjects and the effects of the conditions of the experiment.

The behaviorist position was incorporated into a narrow perspective of experimental techniques, even to the point of insisting upon two variables in defining an experiment (Underwood, 1966). Many areas such as comparative and physiological psychology avoided that error and continued to use all methods of science. Thus, Schneirla (Schneirla and Piel, 1948) investigated ants inductively in the jungles and deductively in the laboratory and co-authored a major comparative text (Maier and Schneirla, 1935) mainly covering research literature excluded from the psychological journals and without restrictive emphasis on any particular techniques. Wever (1949) used inductive and deductive techniques in the comparative study of hearing. Even among the behaviorists there were scientists like Skinner (1956) insisting on the importance of induction. The value of induction is clear from Darwin's ideas on evolution and Einstein's formulation of relativity both concepts produced inductively and not deduced from experiments, except in imagination.

In the 1930's a mostly European school of biology, ethology, began to react against the narrow behaviorists' view. The ethologists insisted on exclusive use of inductive techniques such as field observations of individual organisms, rarely rats, in their natural habitat. Later ethological field procedures were called ethograms. Laboratory techniques were scorned as artificial abstractions removed from natural surroundings. Complex statistical analysis was not needed even in studies of small groups or individuals because of the distinct differences among species.

By 1970, both schools had reached the more universal position of employing all the techniques of science in their work. Many strict behaviorists had adopted a more comparative approach, even though, to some that meant merely adding mice to rat laboratories. Ethologists began to employ laboratory methods. Just when the dispute appeared to have been transcended and synthesis of the schools was expected to allow the values, ideas, and knowledge of both to enhance the total understanding of nature, another group, sociobiology, began to exert influence.

Sociobiology was triggered by Hamilton (1964) who argued that bee worker castes evolved because the altruistic behavior of the workers in raising their sisters was adaptive. The adaptive feature depended on the concept that bee males, or drones, are haploid with all sperm identical. Workers, according to Hamilton, share the same diploid mother and identical genes from a common father, giving workers more genes in common with sisters than workers would have with their own offspring. Workers, thus, contribute more of their genes to the next generation by raising sisters than they would by reproducing themselves.

The unit in evolution is the population, not the individual. Variation in the population, not individual genes, is the central feature of evolution. In the example used by Hamilton that established the foundation for sociobiology, the bees, individual queens fly about 2.5 kilometers from their hives to mate in the air with 6 to 8 drones per flight for typically two flights. Mating with as many as 14 drones is common with all sperm stored together in the queen's spermatheca. In addition, although workers are coded by odor to prevent drifting into alien hives, drones drift freely from hive to hive throughout the spring and summer months. Thus, they mate with unrelated queens from hives distant from the drone's origin. Finally, the newly mated queen returns to her original hive which is vacated by the former queen with her fission swarm. The remaining workers then proceed to tend and raise the next brood, which is the progeny of the new queen and completely unrelated father drones. Therefore, bees are not closely related to hive mates as Hamilton supposed. Rather, genetic traits and even bee races are greatly mixed to provide population variation in each hive, as observations by apiarists confirm.

Subsequent adherents have elaborated sociobiology by applying population concepts like fitness and heritability to individual behavior and individual gene reproduction. Trivers (1972) proposed that gender differences were the result of selection working separately on males and females. Chapter 1 dealt with the impossibility of selection separating sexes in a species. The behavioral patterns, misinterpreted as an individual intent on passing along personal genes, are actually species mechanisms insuring reproductive isolation by means of courtship patterns. Courtship rituals are baroque accumulations of genes, structure, and behavior that creates intra-species attraction and, thereby, prevents interspecies matings. In the theoretical emphasis on the individual, sociobiology is a revision of Lamarckian ideas. Lamarckian principles have had popular appeal, as was indicated in Chapter 1, but the central element, the treatment of the individual as the fundamental unit of evolution, is erroneous.

All Lamarckian interpretations, regardless of popular appeal or use of the language of selection, have true Darwinian alternative explanations based on population principles. For example, Heslop-Harrison in 1926 blamed the predominant occurrence of dark variants of the grey peppered moth (*Biston betularia*) in sooty environments near Birmingham, England, on a "melanogen" consisting of magnesium sulfate or lead nitrate compounds contained in the soot. However, Kettlewell (1959, 1965) demonstrated that Heslop-Harrison's moth population were not genetically homozygous and that the selection factor operating on the moth populations was the vision of bird predators. Thus, in sooty areas birds missed the black varieties and ate the grey. While, in clear environments the grey moths were camouflaged and the black moths were consumed. Natural selection was demonstrated despite controversy.

Selection can not work on the individual or else species could never be cohesive. Each individual would be the source of a divergent species. Individuals can adapt to high altitude or disease but these adaptations are not inherited. Other individuals must adapt on their own. Evolution, thus, is conservative of the characteristics of the species, but not preservative with respect to the individual. A species can sustain individual losses. The fact that the species is the unit of selection will be repeatedly illustrated throughout this book.

The resurgence of Lamarckian ideas at the end of the 19th Century in the political movement called Social Darwinism, as was mentioned in Chapter 1. Social Darwinism used the population principles of evolution wrongly applied to the individual to explain inequities in society. The result created strong philosophical and moral reaction that lead to fundamental misunderstanding of the principles of evolution. Sociobiology's revision of Lamarckian doctrine has triggered similar reaction. Lowontin (1977) and Burian (1978, 1981–82) presented the philosophical criticisms of sociobiology as determinism, meaning fatalism, not scientific determinism as was described in Chapter 1. Creationism is the religious reaction to the moral consequences of the Lamarckian concepts at the core of sociobiology. Life scientists need to be clear in understanding and explaining the principle that the unit in evolution is the species.

Knowledge of organisms and life in general generates, of course, a more accurate and realistic understanding of the behavior and role of the human species in the world. Subsequent Chapters treat humans specifically, while this chapter briefly surveys current knowledge of the evolution of life, not for comprehensive compilation, which would take libraries of books, but for understanding general principles.

In the Beginning

The earth is estimated to be 5 billion years old by radioactive dating of rock (Blum, 1968). Well before the end of the first billion years of the Earth's history, the surface temperature had cooled below 100 °C (212 °F) and thus permitted water to condense from the atmosphere to cover most of the planet's surface, just as the seas do today. The early atmosphere consisted of ammonia (NH_3), methane (CH_4), water (H_2O), carbon dioxide (CO_2), and traces of other materials. In composition, the initial atmosphere of the Earth was similar to that of some planets today. Impinging on that atmosphere was energy in forms such as high frequency ultraviolet light from the sun and lightning discharge from thunder storms. The molecules of the atmosphere were repeatedly broken by the added energy. Other compounds could be formed from reaction of the atmospheric molecules. Large organic substances like amino acids were then produced, for instance glycine (NH_2CH_2COOH) easily forms from ammonia, methane, carbon dioxide, and water. Heat, such as that around hydrothermal vent areas could also have provided enough energy to power the process of amino acid production and the subsequent evolution of cells. In the 1920's and 1930's Oparin and Haldane independently realized the consequences of the conditions of the early Earth's atmosphere and subsequent reactions. Oparin (1968) and Haldane (1954) reviewed decades of work pursing the course of the origin of life. Miller (1953) under the guidance of Urey (Miller and Urey, 1959) began research reproducing the reactions of the Earth's early atmosphere. They used spark discharge in a contained atmosphere of methane, ammonia, hydrogen, and water vapor to produce a variety of organic compounds including amino acids.

The probability of simple molecules recombining into complex forms thus leading to the evolution of life is extremely tiny. However, probability estimates are for instantaneous events. With enough repetitions, even remotely possible events will occur. For example, tossing ten pennies and obtaining all heads is unlikely. The probability is one in $2^{10} = 1024$ chances. However, with many tosses over a long period, the result eventually becomes sure. On the early Earth, over the course of billions of years with countless molecules splitting and recombining, the evolution of life became certain. In Panspermia some argue that the initial materials arrived in meteors or comets with water for the oceans to begin the evolution of life, but the process would be the same.

While knowledge of the origin of life on earth is accumulating, the evolution of the complex and highly organized cellular form of organisms was an even larger step. Indeed, the rest of evolution was a footnote to cellular evolution. Cells came from the evolution of various molecules into the specialized symbiotic structures and organelles of the cell. Thus mitochondria, for example, likely evolved toward specialized existence within autotrophic self-feeding, by photosynthesis, cells from an independent, heterotrophic (feeding on other organic matter) life. Once living processes began, the Earth's atmosphere was fundamentally altered. Carbon and hydrogen were removed while oxygen was liberated. Oxygen formed an ozone layer in the upper atmosphere that protected organisms from further interference by particle and ultraviolet radiation energy. Thereafter, the spontaneous origin of life ceased and evolution held sway. Fox (1970, 1971) as well as Cohen (1970) and Margulis (1971) described the probable process of cellular evolution. Schopf (1983) provides a review of the process of cellular evolution and of current knowledge of the early biosphere.

Blum (1968) offered an explanation of the origin of life and subsequent evolution in terms of the second law of thermodynamics. Given the atmosphere, sea, land, and order, in the uniformly high energy state of the early

Earth, the evolution of life and its species was the result of a continuing chemical reaction. That reaction headed irreversibly toward completion with increasing entropy, and accruing disorder, until all the energy of the original reaction is exhausted. The various forms of life represent phases in the reaction at different energy levels. Although many hold that Blum's theory is non-Darwinian, the formulation does not negate selection. Selection operates through extant conditions to determine the viability of the various phases, which represent the array of species in the reaction. Thus, the thermodynamic explanation offers a powerful conceptual framework for understanding the process of evolution of life in physical and chemical terms.

All organisms remain dependent on solar energy either directly, by photosynthesis, or indirectly in consuming other organisms, or by a combination of both techniques. Humans are an example of a species that employs both approaches. We eat other organisms, but also use ultraviolet light from 2900 to 3200 Angstroms (10^{-10} meters) wavelength to photosynthesize vitamin D in the skin. Vitamin D regulates calcium use in the body. Excess vitamin D leads to calcium deposits and deformation, while a deficiency results in problems like rickets and osteomalacia. Loomis (1967) related the origin of skin color in humans to the requirement of having an optimal balance of vitamin D. Human skin has two pigments, melanin (black) carotene (yellow), in the layer above the vitamin D producing region.

In the tropics where strong sunlight would cause manufacture of too much vitamin D, black skin evolved as a regulator. Dark skin also prevents ultraviolet depletion of folic acid. In more temperate regions, yellow skin appeared. Above 40° North latitude, where human populations spread, not enough sun is present during winter for dark skin to produce enough vitamin D. Therefore, white skin with seasonal tanning by addition of melanin was selected. In the extreme northern areas of Europe even seasonal tanning is absent. When human migration crossed the Bering Strait from Asia into North America and, subsequently, into tropical South America, dark skin secondarily reappeared.

A seeming exception to skin color distribution being based on available ultraviolet sunlight is the Inuit population of northern North America. With the darker skin of their Asian ancestors, Inuits should be unable to exist in

Arctic areas because of insufficient sunlight for synthesizing enough vitamin D. However, Inuit populations are coastal and consume diets high in fish or blubber which are rich in vitamin D. Thus, Inuits behaviorally compensate for the lack of sufficient sunlight and provide proof rather than exception to the concept.

The Evolution of Behavior

Because behavior largely vanishes with the death of an organism, as noted in Chapter 1, tracing the evolution of behavior relies on reconstruction from structural relics, like a jaw with pronounced canines means a predator, and on extrapolation from existing species. However, all existing species are fully modern descendants of more or less closely related common ancestors, and are not themselves ancestral to each other. Thus, the evolution of behavior must be inferred from the behavior of modern forms thought to be similar to that of their predecessors.

Ontogeny is of some value in discovering past evolution. In the last Century ontogeny was thought to recapitulate phylogeny. But embryologic development actually repeats former ontogeny, not previous phylogeny. Ontogeny has been determined by the viability of various subsequent forms of the organisms, like the adult, under selection pressures. The course of ontogeny is frequently altered by events such as mutations or recombinations of genes. A process called neoteny sometimes permits previous larval or immature states to reach sexual function, becoming new species in themselves and bypassing former adult stages. The resulting conglomerate ontogeny is removed from retracing phylogeny. In several instances that will be discussed, including human evolution, neotenous changes led to progressive alterations in the species. However, because all structure and behavior results from transformation of previous structure and behavior, current structure and behavior do not necessarily indicate evolutionary origins. Sociobiology in attributing adaptive significance to current structures and behaviors of individuals may not be correctly interpreting the evolutionary origin of a species characteristic.

The classification system of species, while based on structure, also serves as a guide for tracing behavioral evolution. However, the classification system is itself in flux. Whittaker (1969) proposed a revision based on five kingdoms including monera, protists, plants, fungi, and animals rather than the prior division of plant and animal kingdoms. With basic changes existing at the major level of kingdoms, rearrangement of smaller divisions like phyla and classes is common. Gaps also exist in the knowledge of the behavior of certain phyla. Others like Nematodes, the roundworms, are parasitic, like the hookworm, and are less interesting from a behavioral viewpoint. For convenience, the two kingdom system of plants and animals will be used.

Plants

Plants, although behaving organisms, are little studied for behavior. Plant activity usually occurs at much slower rates than that of animals. Most plant behavior studies deal with tropisms and turgor changes. Watching the grass grow is a popular expression depicting the tediousness of plant behavior observations, even with tools like time-lapse photography. However, Applewhite (1975) reviewed plant behavior experiments. Specific plant and animal interactions will be discussed with appropriate animal phyla. However, since animals are directly or indirectly dependent on plants, the evolution of plants set the stage for subsequent animal evolution. Therefore, a review of the main features of plant evolution follows.

Sometime about 1.5 billion yeas ago, algae appeared. Algae developed an ability to photosynthesize using chlorophyl as a catalyst and were the first autotrophic organisms. Taxonomists usually identify multiple phyla of algae. Bacteria were in existence along with algae and may be related or independent in their evolution. Fungi then appeared as a return to heterotrophic living by relying on dead materials for nutrients and abandoning chlorophyl. Fungi are also categorized into many phyla. Probably, from one of the phyla of algae, the green algae, around 380 million years ago bryophyta, mosses with alternate sexual (haploid) and asexual (diploid) generations, descended. The first vascular plants, the psilophyta, evolved about 300 million years ago. Being multicellular, a vascular system for circulation of water and dissolved food materials was an advantage for psilophyta. Originating from the psilophyta were the lycopods, spore producing club mosses, 280 million years ago and, also, the ferns, which developed seeds, 250 million years ago. The seed is self-contained and independent from parental nutrition. Ferns also evolved independence from water as a medium for fertilization. From the early seed ferns the gymnosperms evolved. The ginko is an early gymnosperm and the conifers, evergreens like pines, spruces, firs, and cedars, are more recent. Gymnosperms developed cones containing unprotected seeds and used the air as the fertilization medium.

Last to appear about 130 million years ago were the angiosperms. Seeds contained in an ovary were an innovation of the angiosperms. Flowers were developed as a means of fertilization. Although other phyla, like fungi, employ alkaloid poisons for defense, the angiosperms used alkaloids extensively for protection to the detriment of some animal species.

The psilophyta, lycopods, ferns, gymnosperms, and angiosperms are all in the same phylum, the tracheophytes. Of the approximately 333,000 species of plants, as many as 285,000 are angiosperms.

Animals

The schema presented in Figure 2-1 will be followed for the animal kingdom including protozoa. The animal phylum, echinoderms, will be considered in its behavioral rather than structural order. Finally, animal phyla with relatively unstudied behavior will be avoided or briefly described.

Protozoa

Protozoa are unicellular organisms of which there are about 45,000 species all of microscopic magnitude and aquatic habitat. They are complicated

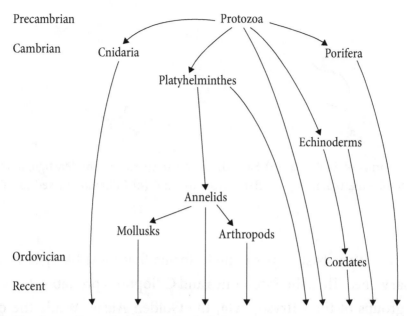

Figure 2-1 The descending order of the animal phyla.

organisms bounded by a membrane, having a nucleus or control center containing the genes, and also various other structures common to all cells. Protozoan origins are in the obscure Precambrian Era over 600 million years ago.

Protozoa are usually grouped according to their locomotive structures. Accordingly, there are three basic types: 1. Sarcodines have no fixed shape, although sometimes they have elaborate shells. Sarcodines have a locomotive pattern marked by projections of their body called pseudopods into which they subsequently stream to a new locale. Allen (1962) has described the streaming activity. *Amoeba* are typical examples. 2. Mastigophora possess a fixed form including specialized structures at the anterior end, like the filament extension, which, when beat, propels the organism forward. Some, like Euglena, have specialized pigmented spots for increased light sensitivity. 3. Ciliophora evidence fixed form with small hair-like structures, or cilia, over their surface. The cilia are coordinated in their movement and propel the organism through the water in a helical pattern. Paramecium are examples having intake openings and a characteristic helical forward movement. Figure 2-2 shows the form of the three basic types of Protozoa. Other categories such as the Sporozoa are composed of parasitic forms of the three basic types.

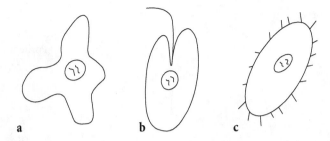

Figure 2-2 Forms of Protozoa. (**a**) Sarcodines; no fixed shape, (**b**) Mastigophora; anterior organization, like the filament-propellent structure, and (**c**) Ciliophora; fixed form and hair-like cilia.

In ancestry, the basic types contain forms that most likely have different evolutionary lines. Thus the Sarcodines and Ciliophora probably descend from different groups of the Chrysophyta, the Golden Algae. While the origin of the Mastigophora is probably from the flagellated photosynthetic Algae, the Pyrrophyta.

Regardless of classification, Protozoa exhibit all the major behaviors of the animal kingdom. They obtain and expend energy by feeding, drinking, and excreting materials. They reproduce, typically by asexual fission, although some, like the Paramecium, anticipate the sexual mechanism by interchanging genetic material through occasional conjugation. They move about their environment. Finally, they respond to chemical, electrical, mechanical, optical, and thermal stimuli. Thus, unicellular organisms, themselves the complex modern descendants of the common ancestors of animal life, illustrate the fact that the basic behavioral repertoire is universal in all life as much as fundamental structures. The first organisms were successful and all subsequent evolution of the various modern forms was the result of accumulated specific advantages from specialization into various particular niches.

In general, Protozoan behavior depends on the strength of the stimulus. They will approach a weak stimulus and withdraw from a strong one. Several weak or repeated stimuli can summate to have the net effect of a stronger stimulus. Schneirla (1959) called the behavior biphasic when the strength of the stimulus determined the direction of motion, approach or withdrawal. He observed that such behavior is universal in animals.

Usually, Protoza exhibit similar reactions to different stimuli with little discrimination or differentiation. However, several phenomena, again universal in animals, indicate an ability to distinguish stimuli. Current activity tends to persist but can be inhibited by a novel strong stimulus. Thus, Amoeba feeding can be interrupted by a strong light. Inhibition demonstrates not only attention, or focusing perceptual devices to stimuli, but selective attention. Selective attention is fundamental to any concept of awareness or consciousness, even though such terms are not understood when applied to Protozoa. The basic process of adaptation also occurs in Protozoa because a prolonged or repeated effective stimulus eventually becomes ineffective. Thus, an Amoeba becomes less sensitive to food after the animal has eaten. Therefore, Protozoan biphasic behavior is greatly modified by both the present and the previous state of the organism.

In modifying behavior with experience organisms use several responses. Habituation is a progressive decrease in response with repeated stimulation. Habituation is typically specific to the stimulus encountered so that a different stimulus could still elicit a response. Jennings (1901, 1906) studied habituation to being prodded by a glass rod or hair in Stentor. He found that initial contractions to first touches required increasing numbers of prods to subsequently elicit a response. Danisch (1921) repeated the work of Jennings with Vorticella using specific weights to calibrate mechanical stimulation. Fatigue was eliminated as an explanation for habituation by eliciting further contractions from strong stimuli. The many modern studies demonstrating habituation were extensively reviewed by Corning and Von Burg (1973).

Several studies deal with the topic of learning in Protozoa. However, situations like sensitization (in which a stimulus to be conditioned already elicits some degree of response before training), pseudo-conditioning (whereby apparent conditioning is due to an extraneous factor rather than the pairing of the experimental stimuli), extinction (a loss of response without reinforcement that, if unknown, could lead to erroneous understanding of results), and spontaneous recovery (the return of response after rest following apparent extinction) could clarify interpretation of learning behavior or even whether terms discovered in investigation of vertebrate behavior are meaningfully applied to Protozoa. Modification of behavior with experience in Protozoa should be interesting in itself without possible inappropriate terminology.

Smith (1908) and later Day and Bentley (1911) found that a Paramecium drawn into a horizontal water-filled capillary tube, too narrow to allow turning, swam repeatedly into the meniscus boundary at one end of the tube, then bent and flipped over. After an average of 8.6 trials the Paramecium flipped with no repeated collisions at the meniscus boundaries. Interpretations of these, and other experiments, claiming modification of behavior with experience in Protozoa as learning have been attacked, because internal chemical changes are apparent and because retention of response is short in duration. Buytendijk (1919) explained the altered behavior of Paramecium as resulting from softening of the cell wall by accumulated carbon dioxide and the concomitant increase in acidity. He found chloroform immersion improved the performance of the paramecium by making them more flexible. Although the controversy is far from settled, such demonstrations do not exclude learning, because whatever learning involves behaviorally must ultimately be reflected by chemical changes in the cell. Applewhite (1975) offered an explanation of habituation in Spirostomum as related to migration of magnesium during acquisition and its active transport back, along with calcium release during dis-habituation. In regard to the duration of the behavioral changes, Day and Bentley (1911) noted some retention of response by the subjects after returning to free swimming for up to 20 minutes. Metalnikow (1912) found Paramecium placed in water with suspended carmine for 24 hours and then allowed to clear vacuoles for up to an hour, accepted carmine less readily. Losina-Losinsky (1931) found the resistance to carmine lasted three days through three generations clearly indicating the profound changes that altered the behavior are retained for relatively long periods in the life of Protozoa.

Much subsequent work has also dealt with learning abilities in many other Protozoa, like the study by Jennings (1923) of Stentor and Vorticella and the study by Mast and Pusch (1924) showing successively less pseudopod extension by Amoeba in the direction of strong focused light. French (1940a, 1940b) used a vertical tube in an experiment similar to those of Smith, and

Day and Bentley. Paramecium had to flip and swim down to escape from the tube. Half of the twenty subjects showed improved performance. Thorpe (1956, 1963) provided an extensive review of the early efforts to demonstrate learning in Protozoa. Gelber (1952) published the first of a series of experiments that described association of a bacteria-coated wire with subsequent approach, compared to lack of approach to an uncoated wire by Paramecium. Jensen (1957a, 1957b) disputed Gelber's findings stating that a bacteria-rich zone had been created by the dipped wire rather than conditioning of the subjects for the wire. Gelber and Jensen began an argument with a series of publications. Their dispute, as well as, other recent work with Protozoa has been summarized by Corning and Von Burg (1973).

Among Protozoa, Mastigophora, considered the most primitive in form, are organized with a gradient from anterior to posterior of sensory structures, like the front tactile propulsive whip, and the forward light sensitive pigmented spot in Euglena. Concentration of sensory structures at the anterior end where stimuli are likely to be encountered is adaptive and has occurred many times in animal evolution.

Origin and Description of Metazoan Features

Whenever a portion of a population becomes reproductively isolated from the rest of the group, the potential for diverging or splitting into a new species exists. Species originate by diverging or splitting. Several factors can contribute to isolation, like geographic separation through geologic change or migration. As long as a group remains one interbreeding population, it is one species with a central tendency or average characteristics. When one group becomes reproductively isolated, it is now free to respond to selection pressures that are different from those of the original population. That group will diverge in a new direction.

Any population is distributed over a range with variations in habitat that result in local characteristics. For example, mammals in the cooler areas tend to be larger than those from warmer regions. In humans, Scandinavians are larger and Mediterraneans are smaller. Size becomes an adjustment to temperature. A cube of side S has a surface area of $6 S^2$ and a volume of S^3. Thus, as the side length S increases, the surface enlarges in proportion to S^2 and internal volume gains by S^3. A small animal, therefore, has a tiny interior volume and a large surface area with which to radiate internal heat in a warm climate. Larger versions of the same animal have less surface compared to internal volume and can conserve body heat better in cool environments. If the population becomes separated, those in the cool area will have different average characteristics and selection pressures from those in the warm regions. Separate species can then evolve.

Divergence, or adaptive radiation, typically occurs early in the history of a species before specialization precludes a wide range of variation. When new species evolve, they tend to diverge rapidly into many environments creating an evolutionary explosion followed by consolidation and specialization into the new niches. Thus, the history of life was a series of spurts. Early life was unicellular but quickly diverged into many multi-cellular phyla. All life, but the Precambrian Protozoa, originated in the Cambrian Period 600 million years ago, except the Chordates, which appeared in Ordovician Period fossils 450 million years ago.

Bonner (1959, 1963, 1967) pointed to the possible origins of multi-cellular forms in his description of the social aggregations of individual slime molds for reproductive advantages. Divergence, once begun, may continue as new species evolve into specialized niches. Parallelism and convergence may also occur. In the phenomena of parallelism closely related lines follow similar paths. The Old and New World porcupines both evolved defensive spines after diverging from a common, but spineless, ancestor. Convergence refers to previously divergent species evolving, by chance, similar structures and behaviors from matching niche selection pressures. The humming bird and humming moth exhibit convergent evolution, as do the wolf and marsupial Tasmanian wolf. Because initial divergence changes the genetic characteristics of species, subsequent convergence is extremely unlikely to reunite separated species.

Comparison of features of the multi-cellular species in future sections leads to the use of descriptive terms like homoplasty, analogy, and homology. Features of two species are homoplastic when they appear alike regardless of function or evolutionary origin. Owl's eyes and the spots on the wings of the Caligo butterfly are homoplastic. The jaw bones of fishes and the middle ear ossicles of mammals are homologous because they share a common evolutionary origin despite the fact that they now function differently and appear dissimilar. However, the hind legs of a bear and a dog are homologous, analogous (because they function the same) and homoplastic. Although structural examples have been used, the same terms may also apply to behavioral features.

Porifera

Porifera are aquatic, mostly marine, organisms with Cambrian Period ancestry. Porifera are composed of an aggregate of similar independent cells which secrete a common, protective housing structure. The approximately 10,000 species of Porifera include the various sponges. Popularly, but wrongly, the term "sponge" refers only to the housing devoid of the hole-occupying cells and to the synthetic imitation. The adult interior cells have a flagellated structure indicating a possible ancestry in colonial Mastigophora. Although, as has been mentioned, ontogeny is not an index of phylogeny. Embryologically, Porifera seem to have features in common with certain colonial Green Algae.

While larval states are usually free swimming, the adults are sessile, bottom-dwelling groups that feed by filtering organic matter from the water. Water is drawn through pores in the colony walls by the beating of the flagellum of the individual choanocytes or collar cells, that line the interior of the colony.

Expelled water travels into the central chamber of the colony and out an upper opening, labeled an osculum. The water pressure generated at the osculum is considerable and creates currents carrying the deoxygenated, food-depleted, and waste-containing water away from the colony. Figure 2-3 depicts a general model Porifera with water flow indicated.

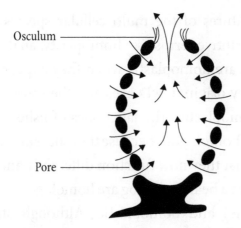

Figure 2-3 Porifera colony with water flow (*arrows*) indicated from pores to osculum opening.

Though individual cells are independent, specialized functions are performed. The collar cells create the current flow. They capture by engulfment (phagocytosis) and digest food. Food is also passed by the collar cells to amoeboid or amoebocyte cells in the colony walls. The amoebocyte cells distribute food to the epidermal cells that line the pores and give support to the colony. The amoebocytes also secrete the housing that is characteristic of the colony. The classification of Porifera is determined by the type of housing or spicules. Calcarea have calcium carbonate housing. Hexactinellida have silicon dioxide spicules. Desmospongiae have horny housing consisting of fibrous protein and are the class used for household cleaning sponges now largely replaced by synthetic "sponges."

Volvox, a colonial flagellate, is hermaphroditic having some central cells that reproduce the colony, even though other cells are quite capable of reproduction. The reproducing cells in Porifera are sexual with some sending sperm swimming away from the colony and others producing and retaining eggs, which, when fertilized, develop into free-swimming larvae. The larvae live among the plankton and later settle to the substrate to become adults.

Muscle cells also appear in the rim of the osculum and control the size of the opening in sphincter fashion. When the colony has not fed, the flagella of the collar cells beat faster increasing water current flow forcing the osculum open. When the colony is food satiated, the flagella beat slows, decreasing the water current flow and the sphincter closes. If probed at one point, the cells of the osculum thus stimulated will contract, mechanically pulling and stimulating adjacent cells. A rapid wave of contraction spreads around the osculum closing the entire sphincter. Such mechanical transmission of information from adjacent cells is called "neuroid" or, perhaps more accurately since nerves are not present, "myoid." Myoid transmission occurs in many animals. The human heart beat consists of waves of myoid contractions stimulated or suppressed by the nervous system. The muscle cells of the sponge osculum also serve as receptor cells. Their response to stimuli is individual but communicated to neighboring cells because of the mechanical connections of the colony housing.

The colonial form of Porifera offers only limited advantage over the single-celled independent life, because increased volume means less feeding surface. As the radius, r, of a sphere grows, the volume increases in proportion to $4/3\pi r^3$ and the surface area enlarges only in relation to $4\pi r^2$. Thus, doubling the radius squares the surface area but cubes the interior volume. Specialization of individual cell function was the behavioral answer to the problem of increasing ratio of colonial volume to surface area for Porifera. The same evolutionary answer was employed in other multi-cellular phyla.

Cnidaria

Cnidaria, also called Coelenterates, are three-layered (outer ectoderm, jelly-like middle layer or mesoderm, and inner endoderm), mostly marine, cylindrical organisms with tentacles at one end around a mouth. The approximately 9000 species exhibit two basic forms. The first is the polyp, which is sessile usually attached to the substrate and has the tentacle and mouth end up. The second is the medusa, which is free swimming, e.g. jellyfish, and like an inverted and flattened polyp. Figure 2-4 illustrates the Cnidaria forms.

The polyp is asexual and reproduces by budding off medusa or other polyps. Some species have polyps with several branches, some of which terminate with tentacles for feeding and others end with reproductive buds. The medusa

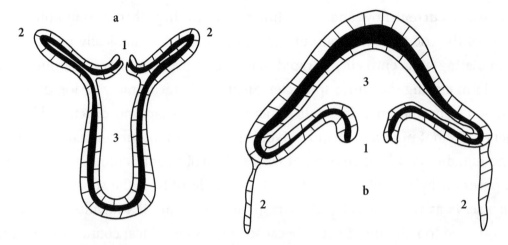

Figure 2-4 Cnidaria forms in cross section: (**a**) polyp and (**b**) medusa. Parts indicated are a (**1**) mouth, (**2**) tentacles, and (**3**) digestive cavity.

is sexual and hermaphroditic exuding gametes into the water to fertilize and develop into ciliated larva, which eventually settle to the substrate to grow into polyps. The mobile medusas are effective in dispersing Cnidaria in the aquatic environment.

While the polyp and medusa stages alternate in the general life cycle of Cnidaria, the three classes vary in the proportions of each of the forms exhibited. Thus, the mostly fresh water Hydrozoa like Obelia show both types. The mainly marine Scyphozoa, like the various jellyfishes, exhibit mostly the medusa phase. The mostly marine Anthozoa, like the various sea anemones appear largely in the polyp form. The popular laboratory demonstration and research Cnidaria, Hydra, is an exceptional Hydrozoa because it exists only in the polyp stage. Anemones were initially mistaken for plants and thus have plant-like names. The corals are Anthozoid polyps which secrete external housing. Cnidaria have evolved from unicellular organisms but the exact route is unclear and depends on whether the polyp or medusa is considered ancestral. Cnidaria originated early in the Cambrian Period.

The tentacles have ciliated cells, which beat to sweep small objects toward the mucus covered tentacle tips. Also present are nematocyst cells, which, when triggered by contact, eject a coiled, venomous stinger helping the carnivorous Cnidaria capture prey or defend themselves. In the polyp stage the tentacles act independently. They flail about until coming into contact with a food object which produces both tactile and chemical stimulation to sensory cells that, in turn, trigger muscle cells. The contracting muscle cells cause the tentacle to draw the food object toward the central mouth and into further entanglement by other tentacles. The mouth has a ciliated epithelium. Normally, the mouth cilia beat to create an excretory current, but food objects brought to the mouth cause reversal of cilia beat, dilation of the mouth, and subsequent flow of food into the gut cavity. The food capturing response is greatly modified by the state of the organism. Thus, unfed individuals more readily accept even nonfood objects, but satiated subjects will respond less actively to food bits and even reject them. The separate sensory and muscle cells are more efficient than the single sensory-effector system in the sponge osculum previously described. In addition, conductor cells organized into an irregular net in the ectoderm of Hydra and both ectoderm and endoderm in many Cnidaria spread stimulation to many effectors.

In the medusa, like that of Obelia, sensors along the rim of the tentacles stimulate the network of nerves to conduct impulses to neighboring tentacle or rim muscles. Medusa sensors include light sensitive ocellus organs and statocyst balance systems. The nervous system has upper and lower interconnected central neural rings. Both the propulsive contraction of the rim in swimming movements and the food gathering action of the autonomous tentacles are coordinated by the nervous system. Strong stimuli will cause more effectors to become activated. Intense stimulation will involve all the tentacles or rim muscles and can cause activity in progress elsewhere to be inhibited or abandoned.

Early investigators felt that the evolution of the nervous system, initially in Cnidaria, represented an opportunity to study an elementary or even primitive nervous system. However, modern animals are not primitive. When a group of organisms evolves a new mechanism, a natural experiment occurs with a variety of forms. Specialization and elaboration proceeds with subsequent groups. Cnidaria, therefore, experimented with many versions of nervous systems. Horridge and MacKay (1962) found symmetrical synapses with vessicles on both sides of the synaptic cleft in the jellyfish, Cyanea. Jha and Mackie (1967) observed both symmetrical and asymmetrical synapses in the medusa of the Hydrozoan, Sarsia. Polarized asymmetrical synapses have been observed in a series of studies by Westfall (1969, 1970) and Westfall et al. (1970) in the Hydrozoa, Gonionemus and *Hydra*, and in the sea anemone, Metridium, in neuromuscular, interneural, and neuronematocystic junctions. Westfall et al (1971) also reported single neurons synapsing with two different effectors, both a nematocyst and an epitheliomuscular cell in Hydra tentacles.

Rushforth (1973) provided a comprehensive review of behavioral modifications in Cnidaria and also (Rushforth et al., 1963, and Rushforth, 1965a, 1967, 1970) described habituation of the contraction response to mechanical stimulation in several species of *Hydra*. The contraction reaction continued to be elicited by light indicating fatigue was not a factor. Rushforth (1965b, 1967, 1970) also demonstrated habituation of the contraction response to light, and to chemical stimulation, and suggested a physiological basis and electrophysiological correlates (Rushforth and Burke, 1971) of habituation in Cnidaria.

Ross (1965) reported association of electrical stimulation of the base, which triggers contraction, with mouth opening behavior in response to squid extract food stimuli in Cnidaria. Seven of 45 sea anemones, Metridium, evidenced mouth opening in response to conditioned electrical shocks. Ross (1965) also described conditioning in the swimming of the sea anemone, Stomphia, which swims to avoid the starfish, Dermasterias, a predator, and contracts to mechanical probing. Ross was able to pair pressure at the base of the organism with chemical stimulation from starfish and found that all of eight sea anemones contracted initially to subsequent starfish stimuli, but quickly resorted to escape by swimming after repeated exposure to starfish.

Echinoderms

The 4000 species of Echinoderms are three-layered marine organisms. Echinoderms originated about 500 million years ago in the Cambrian Period and left abundant fossils because of their limey skeletal plates and spines, from which the phylum name Echinoderm or "spiny skin" derives. Although resembling Cnidaria in some respects, they are quite evolutionarily divergent from them. Figure 2-1 shows the independent but more homologous lineage of Echinoderms and Chordates. The very early Echinoderms were sessile filter-feeders, but modern Echinoderms are mostly mobile predators. Only some of the Crinoid or sea lily class are sessile filter-feeders. They have their limey plates formed into a fused box, from which extend flexible arms that snag small organisms and organic matter. Some sea lilies are free swimming, like coelenterate medusas. Echinoidea or sea urchins also have a rigid body with spines but no arms. They possess pedicellaria or specialized nipping spines. They also have tube feet which are terminals of a water-vascular system that communicates to the external sea water through a sieve plate or madreporite. The tube feet are extended by a series of circular contractile muscles along their length and retracted by longitudinal muscles. Pressure changes in the water-vascular system also affect activity of the tube-feet, especially the flat ends, which can be made concave to form suction cups with powerful grasping ability. The tube feet and also the spines serve a locomotive function. Sea urchins eat sea weed and scavenge. Holothuroidea or sea cucumbers lack arms and spines. The limey plates are small and sparse, giving sea cucumbers a less rigid body. Sea cucumbers feed on small animals and scavenge. Asteroidea or starfishes have a characteristic five-rayed or star shape. The limey plates give the structure stiffness but not rigidity. Spines and pedicellaria are present as well as tube feet. Starfishes eat small invertebrates, like crabs, and, especially, clams, which are pried open with the muscles of the arms that attach by the suction cups of the tube feet. The stomach of the starfish is then extruded out of the central mouth on the underside and into the open clam shell where digestion occurs. Ophiuroidea or brittle stars are similar to starfishes but lack tube feet. Locomotion is achieved by moving the arms which are more slender and flexible than those of the starfishes. Some species of brittle stars have more than five rays. Their diet varies with the species but can include even the tropical-reef corals as does the "crown-of-thorns," Acanthaster planci.

Echinoderms reproduce sexually exuding gametes into the sea to fertilize. Some species of sea urchins and starfish exhibit "brooding" of the eggs, which are covered and protected by the female until they become small adults. In some sea urchin species, the young attach to the body of the adults. However, reproduction by fission is not uncommon, a fact not known to clam fishermen, who do not appreciate competition from starfishes and mistakenly sever the Echinodems into separate arms each capable of regenerating a new adult. Echinoderms have a variety of sensors including specialized cells, particularly concentrated in the mouth, spines, arms, and tube feet. The sensors are sensitive to chemical stimuli, mechanical stimuli, and light. Some have specialized organs like statocyst gravity receptors and ocelli light receptors. Echinoderms are most active at night, concealing themselves in crevices or burrowing during daylight.

The Echinoderm nervous system consists of a central circumoral trunk ringing the mouth region with other trunks radiating into the arms, when present, or body, when not. A nerve-net plexus spreads out from the trunks connecting to sensors, muscles and glands. The nervous system is highly organized to spread the effect of local stimuli, thereby activating remote organs. However, co-ordination of the Echinoderm is mainly mechanical rather than neural. Starfishes, for example, have autonomous arms. Movement is in the direction of one arm which is dominating because of structural features. Figure 2-5 depicts a basic Echinoderm shape with nervous system outlined.

Figure 2-5 Echnicoderm shape. The basic star pattern of Asteroidea echinoderms and the central nervous ring with nerve trunks and connecting nerve net in each ray depicted.

Usually the rays adjacent to the sieve plate are dominant. However, previous stimulation, activity, and the strength of the activating stimulus also determine dominance. Thus, a weak stimulus applied to an arm will elicit a characteristic reflex motion of the adjacent tube feet, leading to movement in the direction of the stimulus. The nervous system, then, activates other tube feet, including those on other arms, which are stimulated by the pulling movement and add to the motion. The tube feet act individually and not in unison, coordinating only their direction of motion. When a strong stimulus is applied, the tube feet retract. The nervous system spreads the excitement causing contraction of tube feet in distal regions but with less strength. Consequently, tube feet of the distal regions are first to extend again and movement away from the strong stimulus results because the distal tube feet take the lead. Central interchange between arms can be prevented by lesions of the circumoral trunk on both sides of an arm. (Single lesions would be circumvented by impulses traveling around the ring in the other direction). Such isolation of an arm results in disunity. If the isolated arm is stimulated simultaneously with another arm in a manner causing motion opposite from the rest of the starfish, fission can result.

Starfishes also exhibit a characteristic "righting" response when inverted. A ray, usually the one being dominant in the immediately preceding behavior, bends its tip around, grips the substrate and pulls the main body over by pivoting on the adjacent rays. Isolating a ray by lesions in the central ring restricts righting ability.

Thus, while the Echinoderms have a larger behavioral capacity than Cnidaria, the similar radial symmetry of both phyla give them a structural organization which permits almost equal dominance of autonomous body parts. Lack of co-ordination in the nervous system further restricts their behavior. Evidence reviewed by Maier and Schneirla (1935) and Thorpe (1956, 1963) for modification of behavior due to experience in Echinodem movement, righting responses, and escape from sleeves placed over arms and pins placed in the angles between rays is inconclusive and inseparable from adaptation phenomena.

Platyhelminthes

The 6000 species of Platyhelminthes or Flatworms are all aquatic, if the aquatic environment provided by parasitized hosts is included. Of the three major classes of Platyhelminthes, two, the Trematoda or flukes like schistosomum (inhabitant of the human bloodstream) and Cestoda or tapeworms like Taenia solium (inhabitant of the human intestine) are comprised entirely of parasitic species. Parasitism is an example of powerful selection pressures favoring short-term adaptive advantage. The long-term effect leads to extinction when the parasite causes reproductive disadvantage for the host. The parasitic species have specialized adaptation to particular hosts and accompanying loss of general motility and behavior. Hermaphroditism is the rule in the phylum and, while individuals are rarely self-fertilizing, they are able to reproduce readily with single or infrequent mating, a necessity for organisms dependent on hosts.

The third class, Turbellaria, is not parasitic. Planarians, a typical Turbellaria, are active organisms resembling Cnidaria, from which Platyhelminthes as a whole have evolved. They have a three-layered body structure with a central alimentary cavity having a mid-ventral tubular mouth, which serves as both entry and exit. Their general appearance, however, is quite different from previously described phyla. Because planarians, and all Platyhelminthes, are bilaterally symmetrical, having mirror image right and left halves when divided vertically from anterior to posterior ends. The planarian shape and general structure seems typical of early Platyhelminthes, which probably originated 500 million years ago in the Cambrian period, but left a poor fossil record because of their soft bodies.

Symmetry has important consequences for the behavior of an organism because structural organization determines the directional orientation of the animal toward the environment. Some organisms like amoeba and snails are asymmetrical and cannot be divided into equal halves along any plane. The Protozoan radiolarians are spherically or universally symmetrical because they are equally halved by any central plane. The Cnidaria and Echinoderms have radial symmetry. They can be divided into mirror halves along only a few radii, like the starfishes. Their structural specialization restricts mobile behavior to the lines of the symmetrical axes, like along the rays of the starfishes. The Platyhelminthes and the phyla, to be considered subsequently, are, in general, bilaterally symmetrical. Bilateral symmetry, with its mirror right and left halves, repeats the trend already demonstrated by the Protozoan Euglena in having an anterior to posterior organizational gradient. Thus, Platyhelminthes are organized about a forward or anterior end. Sensory and neural structures are mainly at the anterior end, where novel environmental stimuli are most likely to occur under normally forward locomotion.

Planarians have an arrow shape with the typical concentration of sensory structures like tactile, chemical, and light sensors at the anterior end. These organisms are virtually transparent. To perceive light, they have evolved specialized, anterior, pigmented receptors called "eyespots," because the sensors are not true eyes but only detect the presence of light. The eyespots give the planarians the false appearance of being cross-eyed to human observers. Beneath the dorsal eyespots is a butterfly-shaped pair of ganglia, which serve to co-ordinate the eyespots and other receptors of the anterior or "head" end. From each of the pair of the butterfly-shaped ganglia, two large neural trunks pass posteriorly with more ganglia occurring along their length. Between these pairs of ganglia, one on each trunk, are cross connections. Peripheral nerves also connect bodily nerve nets to the central nervous system at these ganglia. Thus, the entire planarian nervous system gives the appearance of ladder-like construction. Figure 2-6 shows the planarian structure with eye-spots and the ladder-like nervous system.

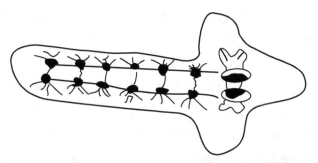

Figure 2-6 Planarian Structure, dorsal view, showing the arrow-shape with eyespots, at the anterior end, above a butterfly-shaped sensory ganglia from which extend the trunks of the ladder-like nervous system. The paired ganglia along the length of the "ladder" nervous system are interconnected and are junctions for the peripheral nerves to connect to and from the body nerve net.

The planarians exhibit three forms of locomotion, described by Maier and Schneirla (1935). They crawl or glide in a layer of mucous by successive local contractions of longitudinal muscles that raise and thrust forward regions of the body. The movement is under the control of local areas of the nervous system, because pieces of planaria are capable of gliding. Also, under strong posterior stimulation, rapid contraction of the longitudinal muscles results in raising of the planarian's center section, which is brought ahead as the center is raised again. The ladder nerves are required to effect the necessary widespread contractions. The result is a humping form of movement. Olmsted (1922) described another form of movement, ditaxic locomotion, in a marine Flatworm, Laptoplana. Waves of contractions travel posteriorly down alternate sides of the worm. Rapid ditaxic contractions result in a "swimming" movement, although Platyhelminthes are not free swimming. They remain in close association with the substrate, except for an ability to crawl inverted on the water-surface film and to lower themselves by their mucous secretions. Dissection experiments prove anterior butterfly ganglia are each originating their side's contraction waves during ditaxic locomotion.

Platyhelminthes generally exhibit the biphasic (Schneirla, 1959) approach to weak stimulation and withdrawal from strong stimuli described previously for Protozoa. Research on the behavior of Platyhelminthes has been extensive. Stasko and Sullivan (1971) reviewed the huge literature of planarian orientation to light. Jacobson (1963, 1965) and McConnell (1966) provide some review of many learning experiments both classical and instrumental. Thompson and McConnell (1955) demonstrated association of planarian contractions to electric shock with light. Succeeding experiments (Jacobson, 1963 and 1965) have confirmed the ability of planarians to learn simple association of stimuli. McConnell, Jacobson and Kimble (1959) began a controversial line of research when they transversely bisected trained planarians and discovered the regenerated anterior and posterior halves both exhibited some retention on retraining. The unspecialized ladder nervous system probably retains the response in both halves. Planarians readily regenerate when divided into halves or even quarters. Spontaneous asexual fission is actually the main mode of reproduction, even though they are hermaphroditic and do exchange gametes. However, Best, Goodman and Pigon (1969) found the anterior end serves to control population density by inhibiting fission when sensory information indicates frequent contact with other members of the species. Corning and John (1961) showed that regeneration in a weak solution of ribonuclease interfered with retention, as would be expected, if planarians have the same long-term ribonucleic acid memory-storage mechanism as do mammals. McConnell, Jacobson, and Humphreies (1961) used the tendency of planarians to consume pieces of other planarians. They found planarians, which fed on pieces of previously trained planarians learned faster than those which ingested naive planarians. Although controversial, the cannibalistic experiments have been widely cited, the implication being that to acquire mathematical skills, one could consume a mathematician or at least the brain of a mathematician. Best (1963), however, dashes the scheme to quick achievement of mathematical skills by explaining that planarians do not digest protein into basic amino acids, as mammals do, but rather, engulf whole proteins and even living cells which can migrate into position within the planarian body. Thus, if the cannibalistic studies prove accurate, and McConnell and Malin (1973) offer strong support, a ready explanation is that whole segments of the neural tissue of the consumed, trained planaria have been incorporated into the nervous system of the cannibal imparting knowledge. Human digestion precludes absorption of whole cells or even whole protein structures.

Best (1963) and Best and Rubinstein (1962) also provided evidence of maze learning in planarians by using a "Y" maze with light and dark alternative choices and return of drained water as the reinforcement. Emotional disturbance was noted, if too much water were drained from the maze making the task life-threatening for the planaria.

Annelids

The Annelids, or segmented worms, consist of some 7000 species in aquatic and terrestrial environments. They probably evolved from the Platyhelminthes in the Cambrian Period but, with only soft bodies, they left a poor fossil record. The lack of a clear fossil record is a major hindrance to tracing the course of their evolution. All of the three major classes are bilaterally symmetrical and have digestive canals with oral and anal ends. They also have circulatory systems with an enlarged anterior "heart" pumping chamber. The original Annelids were probably like the Polychaeta, most popularly known from the marine worm genus, Neresis. The second class, Oligochaeta, is commonly represented by the earthworm genus, *Lumbricus*, which likely evolved from the Polychaeta and in turn gave rise to the third class, Hirundinea, some of whose leech members (Hirudo medicinalis) were in past centuries used in medicine. Annelids are all marked by their segmented structure. All segments are similar but the anterior ones, particularly, show structural specialization in the digestive, circulatory, sensory, nervous, and reproductive systems. The Annelids, except the sexual Polychaeta, are hermaphroditic, like the Platyhelminthes, and are also capable of regenerating after fission. Mating occurs rarely, usually only once, because most Annelids are solitary feeders or scavengers. Earthworms process huge quantities of soil. They digest organic materials, and enrich soil with excretion, as well as aerate the soil by their burrowing. The Annelid ladder nervous system is also similar to that of the Platyhelminthes. In the Annelids, however, each body segment contains one rung of the ladder. The anterior ganglia are also much more specialized, both in structure and in the number and types of neurons. In the anterior segment, several of the ganglia rungs of the ladder have fused into dual, left and right, "Brain" ganglia. These ganglia function primarily to integrate information by selective excitation and inhibition of neural exchange between the varied sensory structures of the anterior of the organism, and the rest of the nervous system. From the dorsal supra-esophageal "brain" ganglia, the trunks of the ladder system pass posteriorly and ventrally around the digestive tract to another pair of ganglia, the sub-esophageal ganglia, also named for their location, and also formed from fused rungs of the ladder system. The sub-esophageal ganglia are mainly motor in function and co-ordinate the musculature and glands of the organism. Posterior to the sub-esophageal

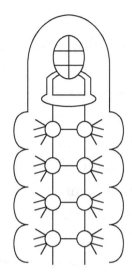

– Brain or Supraoesophageal
  fused ganglia, sensory
– Suboesophageal ganglia, motor

Figure 2-7 Typical anterior Annelid structure and nervous system like that of the earthworm. Each rung of the nervous system is in a separate segment of the animal.

ganglia, the continuing ventral nervous system evidences its ladder structure. Figure 2-7 represents the anterior portion of an Annelid, like the earthworm, and its typical nervous system.

The nervous system of Annelids and subsequent invertebrates also contains a system of usually five "giant" fibers of large diameter. Consequently, giant fibers conduct impulses faster and have a lower threshold of stimulation. The giant fibers course through all segments and seem to have no general role in the action of the ladder nervous system, but may function as a separate, independent, and rapid warning system. The increased number of neurons in the Annelid nervous system also include a specialized kind of neuro-secretory cells capable of changing the activity of other ganglion cells with chemical secretions.

By balancing the tension of the inner longitudinal and outer transverse muscle systems against each other and against bodily fluid pressure, Annelids effect their typical mode of movement. Neresis alternates the contraction of lateral muscles enabling a snake-like "s" crawl. Leeches attach and release alternately anterior and posterior suckers humping the midsection to bring the posterior forward then to extend the anterior in "inch-worm" fashion. Earthworm movement employs the contraction of the transverse muscles forcing forward the anterior end, which grasps the substrate with ventral hooks, or setae, present on each segment. The posterior is then advanced by contraction of the longitudinal muscles. The contraction of the muscles occurs in peristaltic waves from anterior to posterior. Bovard (1918) reviewed experiments demonstrating neural control of the earthworm muscles by separating and pinning front and rear halves of the organism leaving only the neural trunks to connect and transmit information. Peristalsis continued down the entire body. However elaborate the Annelid nervous system, central control does not fully describe Annelid behavior. Friedländer (1894) also cut the nervous system and joined the unpinned halves with thread. The mechanical connection permitted the anterior half to pull on the posterior section whenever the peristaltic contractions arrived at the end of the anterior segment. Forward movement continued in the entire body. Mechanical connection between segments was sufficient to produce myoid conduction (explained in the Porifera) of the peristaltic wave. The redundant neural and myoid systems coordinating earthworm movement provide resistance to damage and illustrate the phenomenon of redundancy common in living organisms. Local autonomy of the ladder nervous structure also influences Annelid behavior. Even behavior apparently requiring the whole organism, like rhythmic feeding and excretory motions, can be performed in isolated body sections.

Jacobson (1963) reviewed recent behavioral experiments with Annelids. Kuenzer (1958) demonstrated independent habituation to a variety of stimuli by earthworms. Local habituation spreads within a stimulated segment and, diminishes into more distant segments. In general, Annelid behavior depends on the strength of previous stimuli, the type of present activity occurring, and the internal states like digestion. Temperature rise also enhances activity because higher body temperature speeds neural conduction. Annelid burrowing behavior has been utilized in learning studies. Copeland (1930), in a classic experiment, conditioned Nereis' response of reaching from its burrow for clam juice, to be elicited by prior light flashes in five trials. The single subject showed an ability to readily reverse its response from light increases to light decreases. Much subsequent work has verified Annelid conditioning. Ratner and Miller (1959a) trained earthworms and also indicated (1959b) removal of the supra-esophageal ganglion had only a small effect on learning and retention. Yerkes (1912) and later Heck (1920) used a "T" maze with several species of earthworms. The choices were light or dark tubes with the wrong choice electrically shocked. Initial learning took between 100 and 200 trials and reversal learning from 50 to 70 trials. Removing the anterior end had little effect on learning, again demonstrating the local autonomy of the ladder nervous system. Robinson (1953) also demonstrated "T" maze learning in the earthworm. Jacobson (1963) considered a wealth of recent studies concluding learning had been proven in a number of Annelid species and noting Yerkes' (1912) observations on the absence of retention in the regenerated anterior end, but also indicating that detailed studies of learning after regeneration are lacking. Arbit (1957) provides evidence that diurnal behavioral cycles affect Annelid learning ability.

Mollusca

The 80,000 Mollusk species, 40,000 of which still survive, are usually grouped into two minor classes, Amphineura and the tooth shells or Scaphopods, and three major classes of concern here: Pelecypods, Gastropods, and Cephalopods. Since Mollusca commonly have shells which serve as skeleton-like levers for the musculature, they left a good fossil record. Amphineura probably appeared first in marine environments in the Cambrian Period, although their fossils are not that old. They evolved from early Annelids or shared with the Annelids a common descent from a segmented group of early Platyhelminthes. Neopilin, a relic species similar to the Ordovician Amphineura, was found living by dredging operations in 1952 at 10,000 feet in the Pacific and still shows internal Annelid-like segmentation. Other Amphineura, the chitons, exhibit segmentation only in their shells. The three major classes rapidly evolved from the Amphineura.

The Amphineura evidence an apparently secondary loss of some Molluscan features, like clear head differentiation. Otherwise, Molluscs have a head with most of the nervous system and sensory organs, a muscular locomotive extension or "foot," a mantle that shields the body and usually secretes a shell, and various viscera like those of Annelids, but also including a kidney-type organ and respiratory gills. The nervous system is very similar to that of the Annelids except for molluscan elaboration of the supra-and sub-esophageal ganglia with consequences for more varied behavior.

Pelecypods, named from their hatchet-shaped foot, are the clams and their relatives. They are bilaterally symmetrical having valves or shells covering their right and left halves, thus the term bi-valves. Pelecypods were particularly populous in the Ordovician Period. Early Pelecypods were mobile. Modern Pelecypods are only marginally mobile. They are mostly marine sessile filter feeders using the Molluscan gill as a sieve. Most clams burrow into the mud or sand substrate like the edible Venus and Mya genera. Others are capable of burrowing into materials ranging from wood like the Teredo or in rock like the Pholas or in coral reefs like the two-meter wide giant clams, Tridacna, of the tropical Pacific. Most evidence an escape response of swimming by clapping their shells together expelling a jet of water and creating locomotion. The scallops, Pecten, are the most accomplished Pelecypod swimmers. Pelecypods are also capable of crawling by extending their foot and pushing against the substrate. They feed by extending their syphon, a modified portion of the ciliated mantle, and extracting organic material from the water.

Pelecypods are highly sensitive to chemical stimuli. They also have a pigmented row, commonly blue, of simple light receptors in the mantle margin between the shells. The Pelecypod nervous system is a modified version of the ladder structure described in the previous worm phyla and discussed under Gastropods below. Figure 2-8 depicts common Pelecypod nervous structures.

Hecht (1929) reviewed his earlier work and concluded that siphon retraction is the result of light impinging on the receptors causing photochemical changes. Morton (1962) examined Pelecypod behavior in response to light, gravity, and chemical stimuli. He found bivalves avoided intense light, crawled down hill, and were sensitive to dissolved substances. While the siphon withdrawal and shell closing response is generally light triggered, the rapid and repeated closing and opening of the shells causing the propulsive escape response is stimulated by chemical traces of such bivalve predators as starfishes. Careful pairing of light and starfish broth stimuli should result in associating the escape response to light. However, very little study of the modification of behavior in Pelecypods has been undertaken.

Gastropods, so named because they appear to crawl on their stomach, are the familiar snails and their relatives. Gastropods, like Pelecypods, were plentiful in the Ordovician Period. In being asymmetrical, Gastropods are exceptional. Some are terrestrial, but in moist areas, rather than aquatic like other Mollusks, and some like the slungs and marine nudibranches lack shells. Terrestrial gastropods have modified mantles that can utilize oxygen from the air in lung fashion. Those without shells usually have other protective adaptations like emitting opaque dyes or incorporating nematocyst stinging cells from Coelenterate prey. Finally, most gastropods are hermaphroditic.

The Gastropod nervous system is a greatly modified ladder structure with the typical Molluscan system of paired cerebral (head), visceral, pedal (foot), and in some species also various pairs of auxiliary ganglia. The pairs of ganglia are the ladder rungs and are connected to other pairs by trunks forming the sides of the ladder. The ladder is not linear but bent with the cerebral ganglia at the apex. Ganglia in the areas for which they are named control the functions of that region. Thus, the pedal (foot) ganglia directs locomotion with local sensory feedback. However, the cerebral (head) ganglia are coordinating centers balancing the local activity of the other ganglia by excitation and inhibition resulting in enlarged capacity for complex behavior. Figure 2-8 represents typical Gastropod neural organization.

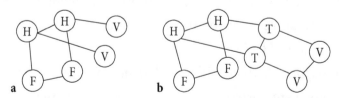

Figure 2-8 Depiction of Pelecypod (**a**) and Gastropod (**b**) Nervous Systems. Paired Ganglia: H = head, V = visceral, F = foot, T = thoracic: each with local control.

Gastropods habituate to a variety of stimuli that initially elicit withdrawal into the shell, but lose effect with repetition of the stimulus. Thorpe (1956, 1963) reviewed habituation and other behavioral work. Thompson (1917) established association between the response of chewing when lettuce contacted the mouth and the stimulus of touching the foot in the aquatic snail, Physa gyrina. The training took 250 pairings and persisted for 96 hours. Conditioning in Gastropods is difficult to establish because the local response must be inhibited. Thus, in the Thompson study, the defensive response of retracting into the shell when the foot was touched had to be inhibited by the cerebral ganglia before the chewing behavior could be evidenced in association with the foot stimulation. Conditioning demonstrates the ability of the Gastropods to centrally modify local responses. Fischel (1931) found the snail, Ampullaria gigas, could learn a "Y" maze in about ten trials, but required daily retraining. Garth and Mitchell (1926) discovered that the terrestrial snail, Rumina decollata, learned a "T" maze thoroughly in about 60 trials and retained the response for 30 days. Hosts of observations on various species of Gastropods and, also, chitons indicate these organism make foraging excursions from fixed sites on rocks to which they return precisely. Pieron (1909) and later Arey and Crozier (1918, 1921) eliminated the possibility that Gastropods marked and followed trails. H. Fischer (1898) witnessed return to the home site along a route different from the outward journey by the limpet, Patella vulgata. Pelseneer (1935) noted returns by the mobile chitons, Chiton cinereus and Chiton tuberculatus, from distances of 1.5 meters. Hewatt (1940) ruled out chance return, and by filing the shells eliminated fitting exact depressions in the rocks. P-H. Fisher (1939, 1950) observed that many Gastropods including the snail, Helix promatia, commonly used for escargot, indicate a complex knowledge of the features of their immediate environment by their homing behavior. A through review of the extensive literature covering the topic of learning in gastropod mollusks is provided by Willows (1973).

Cephalopods include the squids, cuttlefishes, octopuses, and the nautiluses. Early cephalopods were gastropod-like animals very similar to the nautiluses, which, like the chambered nautilus (Nautilus pompilius), inhabit shells built by adding successive growth chambers in an expanding spiral. The cuttlefishes retain a shell, although internally, while the squids have only vestigial shells replaced by an internal cartilaginous skeleton complete with a protective brain case. The octopi have neither shells nor skeletons making them very flexible but unprotected and weak without the substrate to provide leverage. Cuttlefishes, which are exclusively Old World in habitat, and squids are often called decapods because of their ten tentacles. Octopuses have only eight tentacles. In the middle of the array of tentacles is a sharp beak and file-like radula organ for rasping away shells and food. The various small prey are grasped and brought to the beaked mouth by means of dual rows of suction disks along the tentacles.

Cephalopods have elaborate sensory structures, including eyes and statocyst gravity receptors for position sensing. Hochner and Glanzman (2016) describe the evolution of highly diverse forms of behavior in molluscs. Budelmann (1970), Wolff (1970), Schöne and Budelmann (1970), and Budelmann and Wolff (1973) reviewed research and explained the function of the statocysts as that of sensing both linear and angular acceleration. Budelmann, Barber, and West (1973) provided detailed description of the statocyst structures of the octopus (Octopus vulgaris), the cuttlefish (Sepia oficinalis), and the squid (Loligo vulgaris) using scanning electron microscope techniques. Although, cephalopods are not believed to hear, their cutaneous tactile sense provides clues to vibration and water currents. Cephalopods also have excellent vision. The nautiluses have an eye structure like that of a pin-hole camera. While a good functioning eye, it probably provides the poorest view among cephalopods. The cuttlefishes, squids, and octopi have an eye very much like that of the vertebrates in structure and function but of independent evolutionary origin. The convergence of eye evolution between cephalopods and vertebrates caused some difficulty in the history of evolutionary thought, as was mentioned in Chapter 1 in the 1830 debate in the French Academy.

Attendant upon the elaborate cephalopod senses is a highly developed nervous system. Wells (1962a, 1962b) and recently Young (1972) compiled present knowledge of the nervous system of Octopus vulgaris. In the number of nerve cells, the octopus has nearly 2×10^8 compared to almost 2×10^{10} in humans. Nervous system cellular specialization is also complex as is the organization of the neurons. Some 64 distinct lobes have been identified. The lobes in the subesophageal areas are motor in function. The lobes of the supraesophageal region include motor areas, sensory regions, like the optic lobes, and lobes mediating complex functions like the vertical lobes. Ablation studies involving the vertical lobes indicate long and short term memory functions similar to those of mammals. Figure 2-9 depicts the Cephalopod Nervous System.

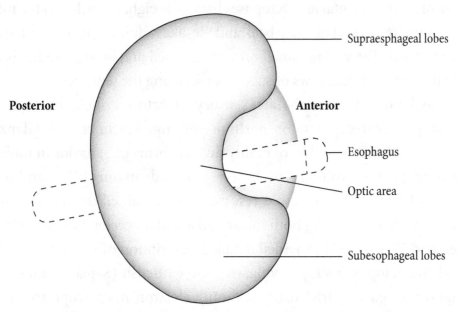

Figure 2-9 The Nervous System of Cephalopods. This is a right side view of the brain of the Octopus vulgaris. The brain surrounds the esophagus with the subesophageal lobes functioning for motor operation, e.g. arms, and the supraesophageal lobes receiving sensory signals and operating conceptual control.

The acute senses and elaborate nervous system underlie a complex behavioral repertoire. Cephalopods have highly individual personalities. They change color, using chromatophores, for camouflage and when excited, clearly showing emotional responses. They exhibit a slow crawl on the substrate using the tentacles, but can swim by expelling a jet of water from the mantle. When frightened "ink" clouds are employed to screen an escape, and in some species irritate the senses of the pursuer. Some shark repellents mimics squid "ink". Although given such names as "devilfish" and species designations like "vulgaris," cephalopods are not likely to attack humans. They prey on small fishes, prawns, crabs, and similar animals. All octopi produce neurotoxic venom from posterior salivary glands, which in some species, like the blue ringed octopus (Hapalochlaena lunulata), can cause problems, if the animals are mishandled. However, these largest of the invertebrates are small creatures. Except for the 18 meter (60 foot) long with .3 meter (1 foot) eye diameter, rarely encountered, deep ocean dwelling, giant squid, Architeuthis or Mesonychoteuthis hamiltoni, preyed upon by sperm whales. Few species are as large as the octopods off western North America, which can span 9 meters (30 feet), tentacle end to tentacle end, with .45 meter (18 inch) bodies like Octopus dofleini.

Octopus mating exemplifies the complexity of cephalopod behavior. In the Winter, the male octopus seeks a mate and leads her to a cavity or nest constructed of rocks. Octopi generally have nests and specific home territories. The male then passes spermatophore sperm bundles into the mantle cavity of the female through the third right tentacle which is hectocotylized. In some species the whole arm is detached and deposited in the female. The male either grows a new arm or dies after mating, depending on the species. Fertilization occurs within the body of the female who then lays the eggs in the Spring. The female does not feed while brooding the eggs for four to eight weeks. Possibly, food might contaminate the eggs. She then blows the new hatchlings free of the nest with a jet of water from her mantle. The young live among plankton until mature in a few months. They then settle to the substrate to pursue adult lives. In some species the young skip the plankton feeding, and in others the new adults remain with and feed on the plankton like the Tremoctopus, Argonauta, and Ocythoe group. Most of the 200,000 hatchlings are consumed by fishes. The female dies shortly after brooding is complete. Octopods have a typical life span of two or three years, and range from littoral zones to over 17,000 feet in the deep ocean like the Cirroteuthis and Eledonella group which includes the blind deep dwelling species Cirrothauna murraye.

Behavioral work with cephalopods has concentrated mostly on shallow dwelling species of octopods with some work on cuttlefish (Wells, 1968). They can make a variety of visual discriminations like squares from triangles, color distinctions, and tactile form perceptions using the sensitive sucker disks. Reinforcement typically consists of shock for wrong choices and crabs for correct responses. Maze learning has also been accomplished. Experiments with octopods are numerous enough to include some dubious techniques like placing a glass barrier between a subject and food. When the octopus repeatedly collides with an invisible barrier, visual ability is blamed. If the barrier is not visible, then only the tactile senses could underlie any response or lack of response. The octopus crevice-seeking is visually triggered by danger stimuli, but the shielding tactile sensations of glass plates will satisfy an octopus, even if not optically concealing. Similarly, jars containing prey cause conflict when brought out of sight under the mantle to the mouth. The jar does not provide the tactile information of prey and is abandoned, until retreat brings the food back into view resulting in the jar being pounced upon again. Eventually the jar will be uncorked and subsequent jar encounters lead to rapid uncorking. Cephalopods, in general are very agile and active predatory organisms. In turn, they are preyed upon by many organisms especially fishes and humans. Various squid species and the octopuses, Eledone maschata, are caught with open pot traps, and along with Octopus vulgaris are dietary delicacies. Cuttlefishes are sought for their cuttlebone. The octopuses are intelligent animals. However, because there is little generational overlap, culture did not evolve. Boycott (1965) and Sanders (1975) describe the large amount of octopus study, including instances of an octopus learning by observing the success of another octopus.

Three times in the history of life, intelligent groups have appeared. One was the Cephalopods, especially octopi. A second was the Hymenoptera, especially the ants. The third was both the Aves and the Mammals. Every time three features were required. One was an elaborate nervous system. Second was well developed senses. The third was dexterity.

Arthropoda

Arthropod means "jointed foot." They are the most numerous of species with as many as 900,000 species or three quarters of all species. The arthropods are ubiquitous throughout the world, except they do not function well below 7.2 °C (45 °F) and therefore, are dormant in winter and absent in arctic climates. Their body consists of an external cuticle, composed of protein and chitin. Essentially, they live in a tubular exoskeleton. Joints in the tube allow flexibility for movement. But the exoskeleton does not expand for growth. Thus, molting is required to enlarge the body. Arthropods are the most numerous in individuals of any phylum, and are more populous than all other phyla combined. There are six major living classes and two extinct, all of which will be briefly reviewed, because the study of any one of them can occupy an entire career. The classes are organized into the subphyla: Trilobites, Chelicerata, and Mandibulata.

Trilobites are all extinct. They were organized with multiple similar segments and represent the ancestors of the modern Arthropods.

Chelicerata, means pincer-like or fang-like mouth parts, and includes the class Eurypterida of extinct sea scorpions, the class Xiphosura or horseshoe crabs, which are not crabs, the class Arachnida, which are the spiders, ticks, mites, scorpions, and relatives, all having eight legs, and the class Pychnogonida or sea spiders. Of the Chelicerata, the spiders are the most well known. All spiders have poisonous venom, although, because of the diseases they carry, ticks are the most dangerous of the Arachnida to humans.

Spiders are all poisonous predators and all are capable of spinning silk fibers from their abdomens. Some are hunters and also ambush prey, mostly insects. The early use of the fiber was as a trail marker to guide the return to the nest. After many journeys, the accumulating fibers probably trapped prey and lead to the evolution of the use of the fibers as a web to catch prey. The web design is species specific and has been employed to examine the effect of drugs on the nervous system by causing alterations in the structure of the web. Spiders can travel by letting out strands in the air currents to attach to another branch and serve as a bridge. People can encounter the web strands while walking outside in the morning. Spiders can also let out strands to function as a parachute when they drop to the ground. Spiders can jump into water and absorb air from bubbles stuck in their body hair. Usually, spiders are solitary and mate by the male tying down the dangerous female. In some species of spiders, the female will tend the young. The water spider (Argyroneta aquatica) uses web to tie together underwater plants and fill the dome with captured air for a lair, the brood, and the young hatchlings.

Mandibulata have antennae and mandibles and include the following four classes. Diplopoda, which are the millipedes. Millipedes are vegetarian consumers of plants and have four legs per segment. Chilopoda, are the centipedes. Centipedes are all poisonous predators and have two legs per segment. Crustacea include the crabs, shrimps, lobsters, and their relatives. They are usually aquatic, have four antennae, and are of commercial importance to humans. Finally, there are the Insects, which include grasshoppers, wasps, bees, ants, flies and many others. All have two antennae and six legs. Insects include at least 800,000 species or two thirds of all species.

Insects show varied behavior and are adapted to almost all environments, except for extreme arctic climates. However, where the warm season is short, insect populations will be intense during the brief warm cycle. They usually have prominent eyes, which are compound. Compound eyes are domes of facets each light sensitive, providing a dot-matrix image. Their range of vision extends from ultraviolet to orange, but not into red (von Frisch, 1971). Thus, red flowers rely on fragrance to attract pollinators. They can also detect polarized light, sensing the position of the sun even on cloudy days. The antennae serve as tactile and chemical sensors. Moths can be detected mates as far as a kilometer (.62 mile) by means of pheromones like bombykol (Schneider, 1974). Fine hair over the body also detect air currents. Some insects have ears on the head like the mosquito. Crickets have ears on their front legs. Many other insects have them on the thorax or abdomen. The ears are chordotonal organs detecting vibration. Insects have four wings, but some have them fused into two double wings. Some, like the diptera, have two wings and two converted into gyroscopic balance organs. Flight may have evolved from beating wings that were originally blood cooling developments.

The nervous system is a modified ladder type with clusters of ganglia in the different segments. The ganglia control local functions in each segment with coordination by the head ganglia. Thus, headless moths can still lay eggs, but can not find the needed substrates. Figure 2-10 depicts the insect nervous system.

Some insects live solitary lives, except for mating. Others are subsocial with adults providing food for larvae. Some are communal, living in composite, but independent nests. When there is cooperation in brood care,

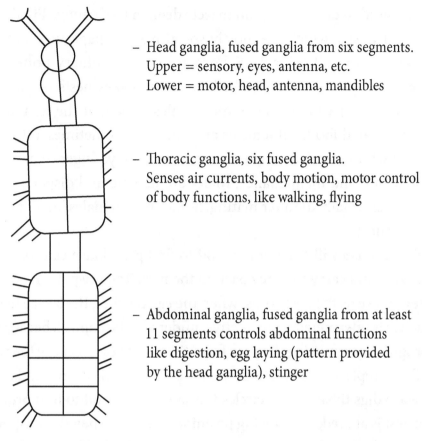

– Head ganglia, fused ganglia from six segments.
 Upper = sensory, eyes, antenna, etc.
 Lower = motor, head, antenna, mandibles

– Thoracic ganglia, six fused ganglia.
 Senses air currents, body motion, motor control
 of body functions, like walking, flying

– Abdominal ganglia, fused ganglia from at least
 11 segments controls abdominal functions
 like digestion, egg laying (pattern provided
 by the head ganglia), stinger

Figure 2-10 Depicts the typical insect nervous system.

that is called quasi-social. When there is division of labor, then the group is semi-social. Finally, when there is overlap of generations, both of which aid in the labor of the colony, then the society is eusocial, as in the truly social insects (Wilson, 1971).

The insect order Hymenoptera, which consists of the wasps, bees, and ants, is a good example of insect behavior. Wasps are predators feeding mainly on other insects. Many wasp species are solitary predators. The velvet ants are actually hairy wingless solitary wasps that hunt on the ground. Sphecid wasps are mostly solitary, digging wasps, like mud daubers, cicada killers, and sand wasps. Chalcid wasps are mainly small and parasitic, laying eggs on other insects. Cynipid wasps inject eggs into plants where the eggs hatch into larvae stimulating the plants to produce galls surrounding the larvae. Ichneumon wasps have long antennae and a long ovipositor, often mistaken for a stinger, used to deposit eggs on insects deep in tree trunks. Finally, there are the vespid wasps, which are mostly social making paper nests of masticated wood, like the yellow jackets, hornets, and the solitary potter wasps, which use mud. The vespid wasp, bellenogaster makes nests of paper cells with many females participating. Some females remain at the nest and feed the larva masticated food. All females are equal, but sometimes a division of labor occurs with older females laying eggs and the younger females doing nest construction and maintenance. Males are haploid and disperse to other nests. The males usually die after mating. Sometimes, females leave the nest to establish new nests.

A sphecid wasp will dig a nest, fly off to find prey, like a caterpillar, sting it to paralyze it and carry the prey back to the nest. The wasp then lays an egg on the prey and seals the nest. Thus, when the egg hatches, the larvae has food. This sequence is interpreted as instinctive, under genetic control, because altering the progress by removing the prey from the nest does not cause the wasp to fly off to find a replacement. Instead, the wasp seals the nest anyway. However, after the wasp digs the nest, she circles the site, and flies off to find prey. If the site of the nest is altered, like moving prominent stones, when the wasp returns with the prey, she goes to the place where the nest should be in relation to the prominent stones. Thus, the wasp, when circling the new nest, memorizes the features in the vicinity of the nest, indicating learning.

Bees are vegetarians feeding on pollen and nectar. Some species of bees are solitary, like the mining types that dig nests in the ground, and like the carpenter bees that burrow into wood. However, most bees are social. Bumble Bees live in colonies of about 100. The colony is founded by a single female by mating in the fall. She survives the winter by digging a hole nearly a meter deep, enlarging the deep end, lining the cavity with moss and sealing into wax a mixture of pollen and honey called "bee bread." She then lays eggs in the the bee bread. The first eggs to hatch provide workers, which are undeveloped females, and a few males. The males are called "drones" because they do not participate in the work of the colony, but disperse throughout the summer to mate in the fall away from their original colony and then die. Late in the season fully developed workers are produced, because there are enough workers to provide the nutrition for full development of females. The females then fly off to mate and repeat the cycle by founding new colonies.

The hive or honey bees, like Apis mellifera, live in colonies of thousands of bees. Lindauer (1961) described communication among bees. The newly mated female, called a "queen" founds a nest and lays eggs every couple of seconds. Newly hatched workers, which are undeveloped females, care for the nest. Males hatch from unfertilized eggs, and do not participate in the work of the colony. They disperse throughout the summer to mate in the fall away from their original colony and then die. Worker activity varies with age. New workers clean the hive and ventilate with their wing beat while stationed near openings. In about six days, the pharyngeal gland matures and secretes a substance for feeding the developing larvae. After fifteen days the pharyngeal atrophies, and a wax secreting gland matures. The worker then participates in nest construction. At twenty-five days the wax secreting gland atrophies and the worker forages for food for the hive until death. If winter arrives, the colony will over winter and continue when the spring returns with warmth. The closely glandular-hormonal sequence of behavior is considered evidence of strict genetic control of behavior. However, bees learn the features of the territory around the hive for about six kilometers or 3.75 miles (Rau and Rau, 1931). The bees forage widely for pollen and nectar. Bees guide by the position of the sun, which, because of their polarized vision, can be followed even on a cloudy day. Then, when ready, they fly directly back to the hive in a straight line called a "bee line." People hunting wild hives for honey, sight along a few bee lines to determine the location of the hive. When the forager returns to the hive with the food, a dance is performed (von Frisch, 1971) to indicate direction to the food relative to the sun. Buzzing indicates the proximity of the food source (Wenner, 1971). Thereafter, many other foragers leave to gather that food. Gould (1986) found bees have a cognitive map of the environment. They will not respond to bee dances indicating a food location in a lake, placed there experimentally on a small boat.

When the queen dies, workers, sensing the loss, will convert developing larvae into queens with a pharyngeal secretion called "royal jelly." If there are no developing larvae, then the colony will dwindle and die. Each colony has an identifying odor, so there is no exchange between hives. Males, however, can and do move from colony to colony. Thus, males are not near their original colony when the fall mating swarms occurs. The females mate with many males in the air and store the sperm. They return to their hives as new queens. The old queen, if still present, will leave with a swarm of workers to build a new colony. Bee keepers watch for these swarms and net them for new nest boxes. Thus, the workers are not especially related to their new "sisters" which are from different males, as was discussed in Chapter 1.

In 1957, Kerr, trying to breed a docile good honey producing variety, imported an African bee variety to Rio Claro, Brazil. The African bees escaped and remained a good honey producer, but are very aggressive. They have spread throughout South and Central America into the southern United States. Aggressive African bees have killed many animals and even some people. The hope remains that the African bees will breed with local bees and become less aggressive.

Bees are the dominant pollinator for many food crops. They have great economic importance to people. In recent years bees have suffered from a disorder called Colony Collapse Disorder, whereby workers do not return to the hive in sufficient numbers to maintain the colony. While, there were reports of this problem as far back as 1906 in the United Kingdom, the disorder has become intense, amounting to almost 30% loss of hives. The cause is uncertain. Suspected factors are mite infestations, immunodeficiencies, loss of habitat, pesticides, and others. Some other insects will perform pollination, like butterflies, but their activity is minor compared to that of bees. Monarch butterflies migrate in the fall from the United States to stay for the winter in Mexico, returning to reproduce in the United States in the spring.

Ants are all social. Wilson (1971) and Hölldobler and Wilson (1990) discuss much of the research on ants. Some ant species are individual foragers guiding by the sun and features of the environment. Other ant species are group foragers and lay chemical trails to food and back to the nest. Files of the group foragers can be viewed on the ground. Ant reproductives fly away from the colony and mate in swarms in the air. The mated "queens" then find locations to start new colonies. The males die after mating. Some ants mate on the ground. Ant colonies can number into the hundreds of thousands and have considerable differentiation. The colonies have unmated males and females with wings, and castes of workers. Some ants have multiple castes with large size workers, especially, defending the colony. Those large size workers are called "soldiers." Schneirla and Piel (1948) described the alternate two to three week encampment and march cycles of army ants as the result of the reproductive process. When new eggs are present along with pupae, the colony has many immobile members and becomes quiescent. When the eggs hatch into larvae and the pupae metamorphose into adults, the colony is mobile again and forages. Some ants, the Attines, cut leaves and raise fungus for food in gardens within the underground nest. Weber (1972) and Hölldobler and Wilson (2011) reviewed the Attines. Weiss (1990) described the ten-year life of an Attine colony. Ritter, Weiss, Norrbom, and Nes (1982) reported that Attines utilize plant sterols in the fungus garden directly without converting them to cholesterol. Thus, the Attines are the first animal found without cholesterol, even in the nervous system. The Attines use an antibiotic to protect the fungus (Oh, et al., 2009).

Some ants like, Formica polyctena, tend aphids, like Lachnus robaris, taking them from the ants' nest to plants and guarding them for the sweet secretion the aphids produce from plant sap. In the evening, the aphids are returned to the nest. Pierce (1985) worked with ants of the genus Iridomyrmex, which tend caterpillars of the Australian mistletoe butterflies, Jalmenus evagoras, which provide a protein secretion. The ants carry the caterpillars up to plants by day and return them to the ant nest at night. Some ants individually use a leaf as a sailboat to ford streams. Weaver ants, like Oecophylla longinoda, use their pupae to spin thread to build the nest in trees. Some ants, of the genus, Polyergus, lack a worker caste and raid other ant, Formica, nests to steal pupae to raise as workers of their own.

Animals that live in tunnels tend to be good maze learners. Schneirla (1934) studied maze learning by ants in a multiple choice point maze. Weiss and Schneirla (1967) demonstrated that maze learning in ants is separated into the run out to food and the return to the nest. Thus, ants following a tortuous route to food do not have to retrace that route to return. Instead the ants can follow orientation by environmental factors, like the sun, to return directly to the nest. Ants are a very complex industrial society in existence long before people.

Termites are another insect that lives in complex colonies. Termites are the most efficient animal at digesting cellulose, by means of symbiotic protozoa in the termites' digestive tract. The protozoa convert cellulose into starch. Thus, termites can eat wood, hollowing out a nest. Carpenter ants also hollow out nests in wood, but do not eat the wood. Both termites and carpenter ants cause major damage to wood structures, like houses. Their role in the environment is to recycle dead trees to prevent disease in the forrest. Termites are not closely related to ants. Termites tend to be white and have straight bodies. Ants are usually black or red and have narrow waists, between the thorax and abdomen. The termite colony can live in ground nests and have mud tunnels up the foundation to the wood of the house. Termites are light sensitive and rarely penetrate wood surfaces. Thus, termite damage can be concealed until the normal-looking wood collapses.

Termite colonies can have five divisions. The first are the winged reproductive males and females, called royals. Termite males participate in the work of the colony. Second are the rudimentary winged males and females, called subroyals. Third are the wingless males and females. Fourth is a worker caste and fifth is a caste of large workers, called soldiers, which defend the colony with large mandibles. Some colonies of the genus Nasutitermes have nasute soldiers, which can shoot a poisonous substance. During spring or fall winged reproductives fly from the colony and find mates on the ground. They dig into the ground and begin a new colony. The King and Queen continue to mate during their life in the colony. In Africa, there are some termites that gather plant material and raise fungus gardens in the nest. If the termites lose the royal caste, for example to an ant invasion, the subroyals can reproduce themselves and the other non-royals. If the subroyals are also lost, the wingless reproductives can produce themselves and the workers and soldiers. Thus, the colony has some resilience. When some termite colonies are invaded by predators, like ants, they can wall off the royals to prevent loss. Arthropods have developed culture, which allows behavior in excess of individual capability.

Arthropod behavior is generally viewed as instinctive, under genetic and hormonal control. Table 2-1 illustrates the comparison between the concepts of Instinct and Learning.

Table 2-1 Comparing the features of Instinct and Learning.

Instinct	versus	Learning
Immediate		Acquired only with effort
Permanent		Can be extinguished
No conflict		Conflict, even with neurosis
Uniform		Individual differences
Fixed		Modifiable

Table 2-1, shows that instinct appears when needed, while learning must be acquired with effort. Instinct is permanent, but learning can be lost. Instinct needs only the eliciting stimulus, but there may be considerable conflict concerning whether a learned response is appropriate in a situation. Instinct is uniform in all the members of a species, while learning can show individual differences, even to the point of a response not being recognizable to others. Finally, instinct is fixed and, thus, adaptive in unchanging situations. For example, the horseshoe crab evolved into the environment of the edge of the ocean, which can alter location, but does not change. Horseshoe crabs have been virtually the same for four hundred million years. Learning, however, is adaptive in providing the ability to adjust to changing situations.

The comparison leads to questions concerning whether a specific behavior is instinctive or learned. However, that comparison is artificial. In reality there is a continuous dimension of modifiability of behavior. Historically, people have labeled responses with low modifiability as instinct and behavior demonstrating wide modifiability as learning. But, there is only a continuous dimension of modifiability in behavior, and not instinct versus learning.

Chordata

The Chordate phylum is the most recent to evolve. Shared characteristics, especially embryonic, with the Echinoderm phylum indicate Chordates shared their evolutionary line with the Echinoderms. The Chordates have a notochord nervous system dorsal to the digestive tract. The Chordate phylum has eight major Vertebrate classes, having a backbone containing the nervous system, in at least 60,000 species. The first four classes are fishes (plural for more than one species). The fifth is the amphibians. The sixth is the reptiles. The seventh is the birds and the eighth is the mammals. Figure 2-11 depicts the major Vertebrate classes of the Chordate phylum.

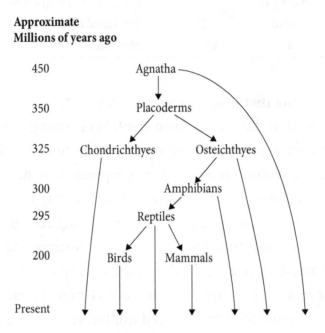

Approximate
Millions of years ago

Figure 2-11 The descent of the major Vertebrate classes of the Chordate phylum.

The first Vertebrate class is the Agnatha, a class of jawless fishes that appeared in the fossil record 420 million years ago. The agnaths may have evolved from sessile filter-feeding animals, like amphioxus, which has a free swimming larval stage. Thus, through the process of neoteny, the agnaths may have evolved by becoming sexually mature in the larval form. However, amphioxus has no ears or eyes, which agnaths do. Agnaths had boney plates, so they left many fossils. They had a cartilaginous skeleton, but no jaw. They were filter feeders. The agnaths became extinct, except for their descendants, the lampreys and hagfishes, which are parasitic or scavenger sucker fishes.

Placoderms evolved from agnaths around 350 million years ago. They were plated fishes with a jaw evolved from the first gill arch. Even though the placoderms were pervasive in the ancient seas, the entire class became extinct, which is rare for an entire class.

Chondrichthyes evolved from early Placoderms in marine areas about 325 million years ago. They have cartilaginous skeletons with vertebrae. Examples are the sharks, rays, and skates. Historically, they were successful and still are active in the oceans. Sharks are carnivores and scavengers. Rays and skates are flat and forage mainly for invertebrates on the sea bottom or filter feed while swimming. Some rays have poisonous spines near their tails for defense. Sharks have denticles covering their body, providing a rough surface. Shark teeth are modified rows of denticles. Lost teeth are replaced by those in the next row.

People are afraid of shark attacks, but attacks are rare, especially considering the millions of people who ocean bathe. Only about 10% of the approximately 300 species of Sharks are known to attack people. Sharks, like all predators, are not mean or evil. They only seek food. Attacks on people, not bleeding in the water, are often bite and swim away attacks, as if they are attempting to drive away a rival predator. A major attempt to develop a shark repellant for downed pilots during World War II used a dye to mimic squid "ink." The black cloud works because sharks avoid uncertain conditions. However, the dye cloud is useless at night or when legs protrude below it. People, on the other hand, kill many sharks for sport and food. The large liver of the shark can be a source of vitamins. In England the "fish and chips" cuisine is sharks. The largest sharks, the whale sharks (Rhincodon typus) and the nurse sharks (Ginglymostoma cirratum), are filter feeders. Nearly all sharks must swim to keep water flowing through their gills. Some sharks, e.g. mako (Isurus oxyrinchus), retain brain, muscle, and visceral heat for faster speed.

Fertilization is internal with the male injecting sperm into the female by means of claspers on the ventral fin. Subsequently, most shark species bear live births. The claspers are the quickest way to identify males apart from females. Sharks have good olfactory organs, vision, a lateral line system of canals along their body to sense low frequency vibration, and ampullae of Lorenzini about the head to sense electrical fields. Overhunting has greatly diminished shark populations.

Osteichthyes are boney fishes, which also evolved from early Placoderms probably in fresh water about 325 million years ago. They spread into oceans and evolved into many niches. There are over 30,000 species most of which are teleosts, which developed an internal bladder as an outgrowth of the pharynx. The bladder originated as an auxiliary source of air bubbled through the gills in poorly oxygenated fresh water areas. In marine environments with more oxygen, the organ became a closed hydrostatic "swim bladder." The teleosts can alter gas pressure in the swim bladder to adjust buoyancy to maintain or change depth without muscular effort. The order ostariophysi have small bones, Weberian ossicles, connected from the swim bladder to the inner ears to enhance hearing. The Weberian ossicles are not related to human ossicles, but function similarly.

Coelacanths were an order of ancient fishes thought to be extinct for 65 million years, until one species was caught in the deep ocean off South Africa in 1938. It was named Latimeria chalumnae after the museum curator who identified it. Finding Latimeria is analogous to finding a dinosaur, also extinct for 65 million years, but less is known of the deep ocean. Another species, Latimeria menadoensis was found near Indonesia in 1998.

Lungfishes can live in warm, muddy low oxygenated fresh water of tropic Australia, South America, and Africa, because they can use the swim bladder to gulp air and bubble it through the gills. When water holes dry up, lungfishes can wriggle across mud to new water holes, using muscular lobes in the pectoral and pelvic fins. The ancestors of amphibians are believed to have emerged onto land by similar abilities.

Fishes have a great range and variety of environments and behavior patterns. Some have internal fertilization with live birth, but most produce large numbers of eggs and sperm to be fertilized in the water. Some remain with the eggs and protect them. In the seahorses, like Hippocampus kuda, the males have belly pouches and provide care for the eggs. Most groupers are hermaphroditic and change sex with age or when there is a paucity of the other sex. Robertson (1972) describes sex changes in a coral-reef fish. Some fishes are luminescent with light emitting photophores in the deep ocean where there is no light.

Some fresh water fishes have modified muscles to provide electrical fields for guidance like Gymnarchus (Lissman, 1963), or for defense or stunning prey, like the electric eel (Electrophorus electricus), which can produce jolts of

600 volts at 50 amps. Other fishes are poisonous, like the stonefish (Synanceia verrucosa) or the red lionfish (Pterois volitans).

Many fishes are vegetarian, while others are predators. Some fishes are mainly solitary, but many others live in large schools using vision and the low frequency sensing (Weiss, 1969, and Weiss and Martini, 1970) of the lateral line organs. Schooling is considered to be a survival mechanism against random predation. However, predation is not random. Sharks can surround a school of fishes to keep them together. Then, the sharks will take turns feeding by passing through the school. Instead, schooling evolved as a tactic to aid reproduction for fishes that exude eggs and sperm into the water.

Migration is commonly exhibited by fishes by daily cycles of moving from shallow to deep water. Fishes are very temperature sensitive. During daylight the shallows can be warm, but cool down at night, causing fishes to seek more stable temperature in deeper water. The movement of food organisms can also be a factor.

Fishes show two types of spawning behavior. One type is catadromous, when fresh water adults travel to the sea to spawn, like eels. The other is anadromous, when adults swim from the sea to spawn in stream headwaters, like salmon.

Aristotle noticed the apparent disappearance of eels at certain times and their reappearance in greater numbers. He supposed the eels spontaneously regenerated in mud. In reality, both European eels (Anguilla anguilla) and American eels (Anguilla rostrata) migrate from fresh water to the Atlantic Ocean to spawn. Eels can live in muddy, stagnant water with low oxygen content, because they have a low metabolism. Eels can crawl over land, especially in wet grass, to seek new water by keeping water in their gills. Oxygen dissolves from the air into the gills to keep eels alive during the land transit. Aristotle's idea was dropped when, in 1896, the Leptocephalus fish was found to actually be the larval eel. Leptocephalus has a leaf-like shape, and is a poor swimmer. It elongates and becomes smaller in a metamorphosis, usually at night, into the elver (rhymes with silver) or young eel. These new eels swim deep at night and largely go unnoticed. The eels live about seven years in fresh water, when they become sexually mature. They also undergo color changes, becoming tasty, and are caught swimming downstream. Eels swim deep and are not easily found at sea. They swim to the Sargasso Sea in the western Atlantic Ocean, where the American eels breed in the western area and the European eels mate in the eastern portion. Thereafter, the Gulf Stream currents distribute the larval eels back to their coasts. The return journey takes a year for the American larvae and about three years for the European larvae. The times are right for the different larval Leptocephalus species to mature into elver and the cycle repeats. The European eels travel as much as 11,290 km (7,000 miles). Eels have been heavily depleted by over fishing, habitat destruction, and chemical pollution.

Salmon breed in the headwaters of streams, mainly in North America. After hatching, the young live in the headwaters where they feed on microorganisms and condition to the mineral content with their taste sense, in the skin, and their odor sense, in their nose. Fish can sense mineral concentrations in as little as one part in a million in water. The young salmon then swim to the sea where they grow to as much as 45 kilograms (100 pounds) in about fifteen years and mature. Salmon then travel to their original stream headwaters, where the females swish gravel into troughs into which they lay many eggs. The males, then, exude sperm into the trough and the females cover the fertilized eggs. Each female will repeat the process with a couple of males. The Atlantic salmon (Salmo salar), after mating, return to the sea to repeat mating for several years. The Atlantic salmon are traversing less distance in slower rivers coming from lower mountains. Thus, they live through repeated mating trips. The Pacific King salmon (Onocorhynchus tshawytsccha), or Coho (O. kisutch), or Sockeye (O. nerka), and others travel as much as 1452 kilometers (900 miles) against swift currents coming from higher mountains. The strenuous journey takes so much energy from the Pacific salmon that they die after one mating trip. Only a few of each type of salmon species are successful in making the full mating trip. Salmon are facing extinction by overfishing, by destruction of access to the headwaters, for example by dams with mainly ineffective passages, and by pollution.

Fishes are good learners with accurate vision. They perform well in discrimination studies, in shuttle boxes, and in water mazes. They respond to positive and negative reinforcement. Sensory capacity, like hearing, which extends from 100 to 4000 Hz, has been determined with conditioning (Weiss, 1966, and Weiss, Hartig, and Strother, 1969).

Fishes have been a major source of food, but are being depleted by river and lake pollution and heavy overfishing by trawlers with nets and long lines. Some attempts have been made to develop aquaculture by raising fishes, like salmon. Several State fish hatcheries, using supplemental oxygen, have been raising fishes for restocking of streams for years. A myth among people who practice catch and release fishing is that fish do not feel the pain of being caught. However, the fish nervous system is well developed and pain is evident in many situations, including electric shock as well as fish hooks. Fishes also do not recover well from fish hook damage.

Amphibians are modern animals. They are not ancestors of current land animals. However, ancient amphibians, perhaps like Ichthyostega with its short legs, were the ancestors of land vertebrates. The land was already occupied by plants and invertebrates, like insects. Thus, the stage was set for vertebrate evolution onto the land. The use of the swim bladder as a lung developed to solve the problem for fishes of low oxygen in drying ponds. That lung enabled fishes to scurry onto land to gulp insects and began the evolution to land living. The combination of ecological and constitutional opportunity led to the rapid occupation of land areas adjacent to fresh water.

Amphibians have altered the basic fish model in many ways. Structurally, the amphibian eye is movement sensitive with no response to stationary food even when starving, especially in frogs. Land movement in air is much faster than that in water. Thus, the fast reflex response to movement in air is required. The ear in larval amphibians, like tadpoles (Weiss, Stuart, and Strother, 1973), being aquatic, is similar to that of fishes. However, for mature behavior on land, frogs, for example, have ossicular columella to connect their ear drums to their inner ears to hear airborne sound. The fins of fishes are converted to legs. Behaviorally, with a nervous system not much elaborated from that of fishes and with a higher premium on speed in air, many actions are locked into reflexes.

There are two major orders of amphibians, the anurans, which include frogs and toads, and the urodils, which include salamanders, newts, and other relatives.

There is also a minor order, the apoda, which are the worm-like caecilians, living in wet tropical leaf debris. The amphibians are not as well studied as the fishes. Salamanders and frogs forage from nests and return. Thus, they are fair maze learners. Conditioning studies have had limited success, because the main amphibian response to novel stimuli is to freeze in position. Camouflage is usually protective when the animal does not move. Mason and Morton (1982) report tiger salamanders, when presented with an odor and intense light, will avoid the odor. Non-conditioning studies using responses, like frogs jumping into water to escape, show a preference for blue, the likely color of water. However, escape training is difficult. Female frogs respond to male mating calls by approaching, but that behavior has not been employed to test hearing. Frogs exhibit a galvanic skin response (GSR). The presence of the GSR in frogs is interesting because they have no sweat glands, which polygraph operators believe are responsible for the GSR. The GSR has been useful to assess frog hearing (Weiss and Strother, 1965), which is similar to that of fishes.

The toad has dry skin and can live and forage further from water, approaching reptilian behavior. However, all amphibians mate in water and the eggs must remain wet. Amphibians, except toads, must keep their skin wet to absorb oxygen. Most amphibians secrete strong toxins in the skin and in the mucus around their eggs. Pollution, chytridiomycosis disease, and loss of habitat are considered responsible for amphibians becoming rare.

Reptiles are thought to have diverged from amphibians, but some argue that they evolved from fishes similar to lungfishes. Reptiles have adaptations favoring the further occupation of the land away from dependance on water. They have dry scaly skin and better lungs, because they do not use supplemental air absorption through their skin, like amphibians do. Fertilization is internal with mating. Reptiles have dry eggs that do not need to be kept wet because they contain an amniotic water sac. Some are viviparous, bearing live young. Many will guard the nests, like alligators, but others, like turtles, do not. Many hibernate in winter. The timber rattlesnake (Crotalus horridus) gathers in large groups, with many other snake species, all hibernating together in a den. There is more development in the nervous system, quantitatively with more nerves, and qualitatively, with olfactory areas of the forebrain starting to evidence enlarged cortical features. In their cortical structure there is better projection of sensory information and more elaborate motor response areas. Adjacent association regions allow increasing flexibility of response. Thus, reptiles have enhanced learning ability compared to amphibians. The reptiles have good vision, sense of smell, tactile sense, and hearing. Wever (1978) documented reptilian hearing range from 100 to 10,000 Hz. Reptiles regulate their body temperature by seeking external heat sources, like the sun or decaying vegetation. In the era of the dinosaurs, the atmosphere was much warmer with ten times the carbon dioxide of today and no polar ice. Thus, the dinosaurs, with their large size, conserved their higher temperature and were functionally warm blooded. They may have had difficulty dissipating internal heat.

The reptiles were a very successful class. They were unopposed in the domination of land areas. Many species appeared and evolved into various niches. The reptiles were experimental land vertebrates in the way that fishes were the experimental aquatic vertebrates. Then, a major dying of those orders of reptiles occurred at the junction of the Permian and Triassic periods, about 250 million years ago. The crisis was not limited to reptiles or even land animals. Many ancient amphibians, placoderms, archaic chondrichthyans and osteichthyans, trilobites, old mollusk and echinoderm species, and brachiopods became extinct or lost dominance. A suspected factor was the breakup of continental masses with much volcanic and earthquake activity. The Appalachian mountains were formed.

Afterward, during the Triassic, Jurassic, and Cretaceous periods, reptiles radiated into new forms, like the dinosaurs that dominated the world. Well known ancient reptiles like the huge herbivorous brontosaurus and like the fierce carnivore tyrannosaurus rex existed. The brontosaurus was so large that it had a spinal ganglion larger than its brain to operate the hind legs, like the rear steering station on a long fire truck. The dinosaurs lasted for about 165 million years until the end of the Cretaceous period, 65 million years ago, then died out. The cause of this great dying is thought to be the impact of a meteor. However, many like Clemens, Archibald, and Hickey (1981) have argued that the extinction occurred gradually over millions of years, not suddenly, and before the meteor impact.

At the time, the flowering plants (angiosperms) were achieving dominance. Many of these plants defend themselves with alkaloid poisons. If the dinosaurs had no taste for bitter, they would have eaten the flowering plants. Thus, the herbivorous dinosaurs would have died, then the carnivores. That progression fits the data of the extinction. This era is also a time when continents were separating again, with much volcanic and earthquake activity. The Rocky mountains were raised. The dying of the dominant reptiles opened the way for new classes, the birds and mammals, which already existed in small forms, to fill the reopened niches.

There are three remaining orders of reptiles. One is the Chelonia, including turtles and tortoises. A second is the Squamata, comprised of lizards, snakes, geckos, iguanas, and tuatara. The third is the Crocodilians, which are the crocodiles, alligators, caimans, and gharials. The tuatara are an ancient lizard-like reptile from more than 225 million years ago, and were previously classified as a separate order, Rhyncocephalia. The two remaining species of tuatara (Sphenodon punctatus) are nocturnal. They eat insects and other small animals, and are only found on about 30 small islands off the coast of the North Island of New Zealand.

Chelonia are the most studied. They have armor in the form of a shell that is part of their skeleton. Some, like tortoises, are terrestrial, others are land and water dwellers, and some are sea turtles with limbs modified into paddles. Turtles can learn patterns (Casteel, 1911), like triangles versus squares, bright versus dark, and others. They can be conditioned. Boycott and Guillery (1962) trained terrapins to avoid food with a particular odor, which was associated with electrical shock. Extinction of learned responses or learning the reverse response is difficult because of strong retention, which results from slow behavior and small number of elements in the nervous system. Turtles forage, mainly for vegetation, about their environment. Thus, they can learn a maze (Yerkes, 1901). However, their slow movement makes maze learning study tedious. The green sea turtle (Chelonia mydas), like other sea turtles, navigates hundreds of miles to specific breeding beaches where mating occurs in the water. The female then climbs the beach at night to dig a hole, into which she lays eggs and covers them. The young hatch in the spring and follow horizon brightness to the ocean. An example migration is from Brazil to the Ascension Islands (Carr, 1965, 1967). Turtles are vanishing from habitat loss, from drowning in fishing nets, from being hunted as food and for their shells, and from food problems, like jellyfish-eating sea turtles misteakingly consuming deadly plastic film waste.

Squamata are not as well studied. Lizards are the most wide-spread of the reptiles. They vary from a few grams to the 100 kilogram (220 pound) Komodo dragon (Varanus komodoensis) of Indonesia. Most are small vegetarians and insect eaters. They are active, depending on their temperature. Geckos and Skinks have separable tails to divert predators. Snakes are descended from lizards with vestigial or absent limbs. The loss of limbs was an adaptation for moving in brush and entering the borrows of prey. There is a myth that snakes do not hear, because there is no apparent external pinna. However, they have ears with an ear canal opening at the surface behind the eyes (Wever, 1978). An external pinna would only rip off in the brush. Snakes also hunt warm-blooded mammals by heat sensing. Vipers have been studied in mazes (Baumann, 1929). They are allowed to bite prey, which is then put in the end box, and the snakes learn the maze to find the food.

Many, but not all, snakes are poisonous. The most well known of the poisonous snakes are those of the cobras, elapid family, and the pit viper family. The United States has both. The coral snakes of the southeast (Micrurus fulvous) and southwest (Micruroides euryxanthus) are cobras and the rattlesnakes, which are widely distributed, are vipers, which have fangs that function like hypodermic needles to inject poison. The largest of the snakes are the boas and pythons of the boid family. They are not poisonous, but kill by constricting coils about prey to prevent breathing. Snakes can open their jaws very wide to swallow prey whole. There are many popular taboos that evidence human fear of snakes. The milk snake (Lampropeltis triangulum) supposedly steals cows milk, despite the fact that they possess no such capability. The mythological hoop snake puts its tail in its mouth and rolls after children. The myths and fears are unfortunate, because they lead to unnecessary killing of snakes. Snakes are important predators. It is probable that bubonic plague, which was spread by rat fleas, was aided by the lack of rat predators in cities and villages, because people exterminated snakes. Some snakes have become damaging and invasive like the Burmese pythons (Python bivittatus) in the Florida Everglades and the brown tree snake (Boiga irregulars) in Guam. Most islands usually do not have native snakes, including Ireland.

Crocodilians populate the tropics of the world. American crocodiles live in salt marches and can be sea going. American alligators are vital to swamp ecology. They hollow areas that fill with water that is necessary In the dry season. Alligators build mounds of vegetable matter in which eggs are laid to incubate. The female alligator will guard the egg mound. Alligators were on the verge of extinction, but, with protection, have recovered throughout the North American southeast.

Much is not known about reptiles. Thus, a lizard on a rock in the sun is said to be sunning. When the animal reaches its preferred body temperature, it shuttles into shade to avoid over heating and returns to the sun when the body cools again (Werner, 2016). Werner (2016) described the life of many desert reptiles. The purpose of sunning is obviously in infrared heat absorption to raise body temperature. However, there are additional possibilities, like visible light regulation of circadian rhythms, or ultraviolet light helping vitamin synthesis. Also, the display could be for territory holding or to attract a mate. Finally, there is the possibility the display helps control population. When there are too many others visible, reproduction may be inhibited.

Birds evolved from dinosaurs about 150 million years ago. Some think the ancestor was the theocodont dinosaurs, others pick the small theropod dinosaurs. The earliest bird is a small crow-sized, feathered fossil, Archaeopteryx lithographica, which retained some reptilian features. Birds are like flying reptiles. The reptilian scale became a feather. The insulating property of feathers enabled maintaining a high body temperature regardless of that of the environment, making birds warm-blooded. Feathers are hollow, strong and light, enabling flight. The back of one feather can hook into the front of another to adjust air flow. Birds also have light, hollow bones with internal cross braces to further flight ability. However, not all birds fly. The largest bird, the ostrich, (Struthio camelus) at 136 kg (300 lbs) is a fast ground runner. Penguins are swimming birds. The smallest bird, which does fly, is the bee hummingbird of

Cuba at 2.8 g (1/10 oz). By the Miocene epoch, 20 million years ago, birds were a dominant class.

Birds have a high body temperature near 43 °C (110 °F), the limit of nerve protein before decomposition. High body temperature requires a high metabolism. Rapid breathing and a four-chambered heart, like mammals, to adequately circulate oxygen, help maintain the high metabolism. High energy consumption for the high metabolism means a need for food. Many birds eat at least half their body weight in food daily. All these features enable flight, which is a high energy behavior. The bird is an efficient flight organism, with a sternum protruding into a keel where the flight muscle attaches. Flight muscle is lighter, with less moisture and stronger for its weight than body muscle. The main flight muscles are the pectoralis major for the wing downstroke and the pectoralis minor for the upstroke. The wing is a modified arm and hand, with arm bones, metacarpals, carpals, and some finger digits fused for strong, rigid movement. The downstroke is for power, but the wing rotates to gain upward movement on the upstroke too, similar to the human hand motion when treading water. Wing beat rate is slower in larger birds. The hummingbird wings beat as high as 90 beats/sec. Sparrows wing beat is 13/sec. Pigeons wing beat is 8/sec. Pelicans have a wing beat rate of 1/sec. Flight speed observations must compensate for the wind currents. Most birds fly about 48 km per hour (30 mph). The woodcock progresses at only 8 km per hour (5 mph). The fastest flyer is the swift, like Tachymarptis melba, which can reach 120 km per hour (75 mph). Hunting birds, like falcons, in a dive can clock at 320 km per hour (200 mph). Hummingbirds, with wing beats as high as 90 per second, are the most agile, being able to hover and even fly backwards. Small and medium birds can take off and land best. In larger birds, the weight per wing area or wing load increases. Larger birds have to run or jump off heights to become air-borne.

The concept of bird flight includes three kinds of behavior. One is true flight, which uses wing flapping to gain or maintain altitude. A second is soaring, which employs rising air currents to gain altitude. Birds can frequently be observed circling without flapping in rising air. The third behavior is gliding, which trades altitude for distance, without flapping. Hunting birds are skilled at combining soaring and gliding to remain aloft for long periods without flapping, thus conserving energy. There is uncertainty concerning the origin of flight. One idea is flight evolved from gliding between trees. Another idea is flight evolved from running and jumping up to catch insects.

People have always interacted with birds. Birds have been a source of food and fascination, especially for their ability to fly. Many individuals and groups observe bird behavior. Bird food is a multi-million dollar industry.

One of the behaviors that intrigue people is homing, the behavior of having a home base. Birds often wander far in search of food, then return to their nest. The common pigeon or rock dove (Columba livia) is the most studied. The pigeon has good flying ability, usually between 33 and 100 km/hour (20 to 60 mph), and has been domesticated and bred for its homing ability. Until the advent of electronic communication, the homing pigeon was a valuable means of sending messages. The pigeon can return from over 167 km (100 miles) with high probability. The journey can take more than one day, indicating feeding and persistence to return. To be used to deliver messages, pigeons are trained by progressively further releases to enable landmark learning. For reliable delivery, several are released, each with the information. Stories of messages delivered by pigeons at critical times in history abound. The return time is more than double for double the distance into strange territory. Probably, when released, pigeons fly in expanding spirals until familiar territory is reached. Pigeons taken outward in rotating cages, still returned home, hence they were not retracing the outgoing course. The height necessary to see home from large distance is likely too great. Most birds fly below 600 m (2000 feet) and are rare at 900 m (3000 feet). The need for food and water necessitates staying near the ground, but birds can home over water. The Manx Shearwater, displaced by people to Boston, returned to its nest burrow in Wales in 12.5 days (Mazzeo, 1953).

Birds are usually active in daytime, when sight can be employed. Flying through branches at night is dangerous. Night is for sleeping. Even night hunters, like owls, move to a low perch at dusk and hunt from the perch to the ground using hearing. Bird hearing is slightly less sensitive than humans (Dooling, 1982) and extends from 200 to 10,000 Hz, except echolocating birds, like oilbirds (Steatornis caripensis), have frequency ranges out to 15,000 Hz (Konishi and Knudsen, 1979). However, vision is acute, especially in hunting birds, which see better from farther distances than humans.

Migration is different from homing. Birds migrate in the fall from the breeding range in the north to an over-wintering area in the south and return in the spring. In addition, young birds on their first migration are flying to places they have never been. The young birds are not just following the older birds, because in some birds young and old take different routes or the young leave before the older birds. Migration has always fascinated people. The phenomena is mentioned in ancient texts, like the Bible (Exodus, Chapter 16, verses 12 & 13), which states Quail arrived in the evening to feed the wandering Israelites. Manna was a vegetable gathered on the ground in the morning.

The migration of many birds is spectacular. The Arctic tern (Sterna paradisaea) is the most dramatic. The Arctic tern breeds in the Arctic in summer. When winter approaches in the Arctic, the Arctic tern migrates to the Antarctic, where there is summer. The trip is over 20,150 km (12,500 miles) each way. It is unknown how such a fantastic journey developed. The Golden Plover (Pluvialis dominica) migrates from Alaska to Argentina a distance of 13,000 km (8000 miles) and back. The Blackpoll Warbler (Dendroica striata) migrates from Alaska to central South America traveling 7500 km (4500 miles), taking 86 hours and being spotted by pilots at 6300 m (21,000 ft). The tiny Ruby-throated humming bird (Archilochus colubris) migrates from the eastern United States to Central America nonstop across the Gulf of Mexico. However, the American Robin (Turdus migratorius) population of eastern Canada migrates in anticipation of winter to New Jersey, where it over winters, while the New Jersey Robins migrate to North Carolina for the winter. That migration indicates the behavior is a whole population phenomena, because, if New Jersey robins remained, then Canadian robins would have nowhere to go. In summer Robins eat worms, but in winter they eat berries. Canadian geese (Branta canadensis) fly in flocks with one goose leading. The others follow, riding the leader's wake in a V-shaped formation to conserve energy. The geese take turns at the lead. The high need for food means birds must flee the breeding grounds ahead of winter for better foraging in the south, where they do not breed. Environmental factors like cold, day-length, celestial patterns, and favorable flight weather conditions trigger the fall migration by reducing feeding time. However, many of these factors would not be present in the south to send the birds back north. The migration patterns, both departure and arrival, are fairly precisely timed to the time of year.

The birds follow particular flyways along major geological features. In North America there are four flyways. One is along the eastern coast. Another matches the Mississippi River. A third traces the eastern edge of the Rocky Mountains. The fourth follows the western coast. All flyways are along areas of abundant food and water and terminate in warmer areas. Human development along a flyway can break the path by opening gaps with too much separation and no place for the birds to land and feed. People observing birds take up positions along the flyways at the appropriate season. Other continents have similar flyways. Some European flyways are along the Atlantic coast, down Italy, and around the eastern Mediterranean. The southern ends of European flyways are in tropical Africa. In addition to geological features, birds can guide by the sun and star patterns (Sauer, 1971). Some birds are sensitive to polarized and ultraviolet light (Kreithen and Eisner, 1978), to the magnetic fields of the Earth (Wiltschko and Wiltschko, 1995), to barometric pressure (Kreithen, and Keeton, 1974), and to the infra-sound of ocean surf rumbling (Kreithen, and Quine, 1979). Migration requires enormous energy using all the fat reserves of the birds. Estimates of over 10 billion birds migrate world wide, but the losses in the returning bird numbers can be as high as 40%. Loss factors include fatigue, habit destruction, flyway interruption, chemicals like insecticides, predators including those newly introduced to the territory, hunting, and poaching.

In the spring birds return to the same area, which they left in the fall, even to the same nest. Older birds will hold their territory until the younger birds take over. Then, the older birds seek new locations. Song birds, which includes many birds, hold territory proportional to their food requirements. For some that can mean a fifth hectare (.5 acre), others will hold .4 hectare (one acre). Owls can hold 10 hectares (40 acres). Sea birds do not need territory for food, since feeding is accomplished at sea. Therefore, only a small nest area is necessary, but nesting space on shore can be tight. Thus, nesting can occur in large colonies, called rookeries. Usually, the males arrive before the females. Some birds mate for life, which means the loss of a mate ends the reproductive life of the remaining bird. Others will accept new mates, if the original mate is lost. The male sings in a place visible throughout the territory, from the bird viewpoint. The singing holds territory and attracts a mate. If another bird enters the territory, the holder will sing louder. If the intruder does not leave, the holder will approach. If the intruder still does not leave, there will be a skirmish with the winner taking the territory. Usually, the intruder is uncomfortable and leaves before a confrontation. Large birds and small birds can share territory, because they are separate species with different feeding habits. The presence of a bird feeder will cause many birds to invade a territory, but the abundance of food makes the defense of the territory less necessary. Territory holding functionally distributes the bird population over the breeding area.

When the female arrives, nest building begins. In some birds, both mates build the nest. With other species, the female builds the nest and the male guards the nest to prevent theft of the material. Fertilization is internal and the female lays a few eggs. In many species both mates incubate the eggs with brooding spots that become warm and moist. For a few species only the female broods, while the male begins foraging for both. If an egg rolls away from a ground nest, the parent bird will retrieve the egg by walking up to the egg, hooking its beak over the egg, rolling the egg backward, then backing up and repeating the process until the egg is returned. If an egg is removed from the nest, especially when the nest holder is away, the parent will accept another egg upon return. Some species will remove the eggs of others, replacing them with their own. Farmers, hand collecting eggs, use glass eggs to replace harvested chicken eggs.

When the eggs hatch, singing becomes less frequent because of foraging requirements for the young. For most birds, about a third of the nestlings die, because the adults can not keep pace with the feeding requirements. The stronger young will take a lot of food, even pushing the weaker siblings out of the nest. In rookeries, the young can wander into another nest. If the owner is present, the chick will be pecked and injured or killed. if the owner is absent, the chick will likely be accepted upon return of the nest holder. After a few weeks, the young will leave the nest and forage on their own, but avoid competing with the older birds until the young return from their first migration. The parents, in most species will attempt to raise another brood, if there is time until fall.

Many birds have a social order established by pecking, hence it is called a pecking order. Barnyard chickens have been observed (Guhl, 1956) forming a complex order as chicks, but by 36 weeks the order is linear. The top chicken gets first choice of food. In chickens, the males and females have separate orders. In other species, there can be a mixed order, but a top female is rare and must yield her position for mating. The social order is established by aggression, but serves to limit fighting as long as there are no challenges among the hierarchy.

Birds are excellent learners in many different task and problem situations. They can use tools. For example, seagulls will carry clams aloft and drop them on rocks to open the clams. In 1948 there was a movie called "Bill and Coo" directed by Dean Riesner. The cast was parakeets as the citizens of Chirpendale, who are terrorized by an evil crow, but saved by Bill, a parakeet. The entire cast was birds performing incredible acting behavior.

There are some invasive bird species. Eugene Schieffelin, Chairman of The Acclimatization Society, which sought to import animals and plants thought desirable, released 60 European starlings (Sturnus vulgaris) in 1890 into Central Park of New York City. He released another 40 in 1891, supposedly, to bring all the birds mentioned in Shakespeare to the United States. The starlings proliferated. He released other birds also, but, except for the house sparrow (Passer domesticus), none survived. The starlings reached the west coast of North America by 1950 and ranged from Alaska to Mexico. Estimates of over 200 million now exist (Todd, 2001, and Mirsky, 2008). The starlings are migratory in Europe, but only gather in large flocks without migrating in North America. They breed in holes and out-compete native hole-nesters like woodpeckers and bluebirds. Starlings are responsible for one billion dollars in crop damages. They also spread diseases including histoplasmosis, toxoplasmosis, and Newcastle disease. The last mentioned affects poultry. On October 4, 1960, a flock of starlings brought down Eastern Air Line Flight 375, a Lockneed Electra, taking off at Logan Airport in Boston. That crash killed 62 of the 72 people on board.

Currently, there are an estimated 10,000 species of birds. At least one third are endangered or near extinction. Even before the current count, hundreds of species have become extinct. The causes are destruction of habitat, over-hunting or collecting, predators, newly invasive predators, and chemical pollution like pesticides and fertilizer.

Mammals evolved from therapsid dinosaurs about 100 million years ago, making mammals fairly recent animals. The early mammals were small and probably nocturnal to avoid becoming prey for dinosaurs. When the dinosaurs became extinct, mammals, along with birds, filled the vacant niches. Currently, there are some 26 orders of mammals, which range in size from tiny mice to the blue whale (Balaenoptera musculus), the largest animal ever to inhabit the planet. Mammals have been very successful with 4785 species. Some of the major orders will be reviewed. All mammals provide milk to feed the young. Only the Monotreme order, which are the platypus (Ornithorhynchus anatinus) and the echidnas, lay eggs, but still provide milk to their young. The Monotremes reside in Australia and New Guinea. All other mammals have live born young. The Marsupial order, however, bear extremely immature young weighing only a few milligrams. The young usually develop in a pouch or fold of skin where they feed on their mother's milk. The most well-known are the common opossum (Didelphis marsupialis), the kangaroos, and the koala (Phascolarctos cinereus). The various species of opossums appear in the Americas. The other Marsupials inhabit Australia and New Guinea, but have been introduced to New Zealand, Hawaii, and Britain. The common opossum is famous for playing dead when attacked. They carry the young in a nursing pouch for two months, then the young ride their mother's back for about four months. The red kangaroo (Macropus rufus) males are known for boxing with their front limbs and kicking with their powerful hind legs. The western gray kangaroo (Macropus fuliginosus) forms mobs, or social groups, in families. Some Marsupials, like the Tasmanian devil (Sarcophilus harrisii), are meat eaters. The rest of mammals are placental, carrying an internal fetus until birth.

The order Chiroptera, bats, are the only flying mammals. Some lemurs and squirrels labeled as flying can only glide between branches. Bats are very agile flyers with flexible wings composed mainly of unfused arm and hand bones with skin covering. They are not strong flyers, like birds. The largest group, in size, are the fruit bats, or flying foxes, which are mainly fruit eating, and can have wing spans of up to two meters (80 inches). They find food with vision. The smaller bats, or microbats, can have a wing span as small as 15 cm (6 inches). Microbats mainly eat insects and hunt by echolocation with high-pitched clicks and sharp hearing (Brigham, et al., 2004; Simmons and Stein, 1980). Three tropical species are called vampire bats, because they bite and lap the blood of animals, mainly cattle. Bats are social, nesting in colonies, and flying mostly at night.

Some mammal orders are aquatic. The order Cetacea includes whales, dolphins, and porpoises. The large whales, are called Mysticeti or baleen whales, because they use baleen sieves to collect plankton. The smaller whales, called Odontocetes or toothed whales, are whales, dolphins, and porpoises. The toothed whales hunt fishes and aquatic mammals. Whales are social and intelligent. They communicate vocally and employ echolocation (Au, et al., 2000) which is useful in their low-light environment. The baleen whales have been hunted to near extinction for oil, baleen, and meat. The use of less expensive petroleum oil saved the whales from extinction. Pinnipeds are another aquatic order and include the seals, walruses, and sea lions. They are all predators, mainly on fishes and invertebrates. They spend much time in the water, but breed and rest in social colonies on land.

Three orders of mammals are herbivores. All these orders walk on their toes, some formed into hooves. The Perissodactyla are odd-toed and include horses, zebras, donkeys (All three are close relatives and can produce hybrids.), tapirs, and rhinoceroses. The zebras are the most camouflaged with black and white vertical stripes. The vertical stripes in the tall grass and the moonlight hide the zebras at their most vulnerable time from nocturnal predators. The Artiodactyla are even-toed and include pigs, sheep, hippopotami, camelids, deer, giraffes, buffaloes, bison, cattle, goats, gazelles, llamas, and antelopes. The third order, Proboscidea, is the elephants, both African and Asian. All of these orders dwell in herds and rely on fleet escape. Some can use horns or tusks for last resort defense. The large size and bulk of elephants, rhinoceroses, and hippopotami also discourages predators. The herds are mostly groups of females with offspring and a dominant male. Elephants live in matriarchal herds with an older female leader, who knows where seasonal food and water resources are located. The large pinna of elephants can collect very low frequency sound, which carries over long distance. Many of the herd animals, like the American bison and the elephant have been hunted to near extinction. Tusk ivory is the main reason whole herds of elephants are slaughtered.

The order Carnivora includes wolves, dogs, coyotes, foxes, cats, hyenas, bears, pandas, raccoons, weasels, otters, badgers, skunks, jackals, civets, mongooses, and others. Almost all are predators with varied hunting techniques. Some are omnivores. Pandas are vegetarian, with the giant panda (Ailuropoda melanoleuca) eating only bamboo. The sea otter (Enhydra lutris) spends most of its life in the costal ocean diving as much as 100m (330 ft) to collect clams and a rock. Returning to the surface, they float on their backs with the rock and clams on their bellies. They use the rock as a tool to break and eat each clam in turn. Wolves and dogs are pack hunters, running after and overwhelming prey. They communicate many emotional states through posture, yelps, barks, howls, scent, and facial expressions. The cats have been very successful in various forms from large versions hunting big prey to the smallest cats (Felis catus) domesticated in Egypt and in Asia, probably for rodent control. Cats are mainly solitary hunters, except lions (Panthera leo), which live in family groups, or prides, and feral domestic cats, which can live in colonies. The lion prides have a dominant male and several females with cubs, as well as, young males and females. The adult females cooperate to kill prey. Then, the males, afterward the females, and finally the young eat. The male holds the territory, engaging in fighting only when another pride or lone male enters the area. Then, the dominant males will fight with the winner taking the territory and most of the females. The losing male and his remaining pride seek a lesser area. The winning dominant male will often kill the cubs of the previous male to replace them with his own offspring. Young males can leave on their own and seek to establish their own pride elsewhere. Pride leadership requires experience as well as strength. Captive lions in natural park settings, do not form prides without older experienced males. That was the experience of the Yester Lion Country Safari Park in California. The eventual leader of their pride was a senior male, named Frasier, who sired 35 cubs and became a celebrated symbol for senior human groups and a movie (www.yerterland.com/lioncountry.htlm). While lions are plains animals, tigers are solitary jungle hunters. Their ranges do not overlap, but they are closely related enough to mate and produce hybrids in captivity.

People keep dogs and cats as pets and argue which is better. They argue that dogs can be trained and are obedient, versus cats, which use a litter box and wash themselves. In reality neither is better. Each has evolved into different niches. Dogs are pack hunters running after prey. The lead dog engineers the kill and shares food. Thus, a human sharing food becomes the leader of the pack and is followed. Cats, on the other hand, are solitary hunters hiding their waste to conceal their presence from prey. Cat washing occurs after eating, not before, and serves to cover their body odor with the food odor, further creating concealment. Cats are also blamed for killing birds. Rarely, can a healthy bird be caught. Thus, cats eliminate the ill and infirm and prevent an epidemic among the birds. The main prey of cats, however, are rodents and, where available, small reptiles, amphibians and insects (Arnold, 2015). Cats are valuable in reducing the rodent population. As was already stated in the discussion of birds, the main cause of bird loss is the destruction of their habitat by humans. Cats have high body temperature, providing fast responses. They rest frequently to conserve energy. The domestic cat has become commensal with humans.

The Rodent order is huge with about 3000 species ranging from tiny mice to the pig size South American capybara (Hydrohoeris hydrochaeris). Almost half of all mammal species are rodents. The order includes mice, rats, voles, beavers, porcupines, guinea pigs, hamsters, jerboas, squirrels, chinchillas, gophers and others. Most are vegetarian, but others are omnivorous opportunists. Attempts to categorize the 33 families have provided two subgroups, the globally widespread Sclurognathi or squirrel-jawed, and the mostly southern hemisphere Hystricognathi or porcupine-jawed. The Sclurognathi include all squirrels, beavers, and rat or mouselike rodents. The Hystricognathi are the guinea pigs, porcupines, chinchillas, capybara and others. All rodents have an upper and a lower pair of incisors that grow continuously and must be trimmed by gnawing. The rodents are very pervasive. Most are small and nocturnal, with high reproductive rates. The common house mouse (Mus musculus) has followed humans world wide, even into the Antarctic. Mice, in general, threaten human food supply by eating or contaminating crops and stores. Rats live in large colonies and can devastate human food stores and crops, as well as spread disease. The European or brown rat (Rattus norvegicus) and the black rat (Rattus rattus) have caused the death of millions of people. During the middle Ages, the brown and black rats killed as much as a third to half of the European population by spreading bubonic plague with their fleas. The brown and black rats have spread throughout human habitation globally by stowing away on shipping. These rats are prevalent in cities where there are few predators.

The North American beaver (Castor canadensis) and the Eurasian beaver (Castor fiber) mate for life and live in families. They eat bark and gnaw through small trees. The cut trees and branches are dragged into nearby streams where beavers built stick and mud domed dams. The beavers live in the hollow dam lodges, with underwater entrances. The beaver dams block streams and accumulate pond water that becomes a vital supply for other wildlife. Beavers have been heavily hunted for their fur. In France, the beavers were almost extinct. Then, the beavers were protected by the French government and they increased in numbers. For the first time in about 300 years, French beavers in southern France built their dam-lodges. Since no beavers remained from past generations, some scientists claimed proof that beaver dam building was genetically programmed. However, given the structural and behavioral capability of beavers, when they learn to build dams the dams they build will be beaver lodges, not hydroelectric power dams.

The order Lagomorpha includes rabbits, hares, and pikas. These animals have evolved into similar niches to those of rodents and are often confused with rodents. However, Lagomorpha have different features like two pairs of both upper and lower incisors, for a total of four in each jaw. Lagomorphs tend to be nocturnal and use eyes and large ears to sense predators. Their long hind legs allow the animals to hop very rapidly away from danger. They are herbivorous and have high reproductive rates.

The Primate order has two suborders, the Strepsirrhini and the Haplorrhini. All primates have relatively large brains and exhibit very flexible behavior. They are arboreal with some ground excursions. They have good olfactory senses, vision, and hearing. They range from small family groups to large packs. An opposable thumb, evolved for grasping branches, is a hallmark. The main concentration of primates is in the tropical regions of Africa, Asia, and the Americas. The Strepsirrhini are the lemurs, lorises, bushbabies, galagos, tarsiers and relatives. They are mostly nocturnal vegetarians or insect eaters with adaptations for the dark, like no color vision, to be more sensitive in low-level light. Some, like the lemurs of Madagascar can be cathemeral or active on the ground in daylight, at times, as well as night. The Haplorrhini are the new world monkeys, the old world monkeys, and the apes. They evolved for day-time tree living with stereoscopic vision and color sensitivity for discerning the next branch. Tree life is diurnal, because moving through the trees at night is dangerous. The new world monkeys are platyrrhine, having side-facing nostrils and prehensile tails. They range from tropical rain forests of South America to as far north as Mexico. They include the howler, spider, woolly, and squirrel monkeys, along with the marmosets and other relatives. The old world monkeys are catarrhine, having downward-facing nostrils and lacking prehensile tails. They are distributed across forested regions of Africa and Asia. Most are arboreal, but baboons are terrestrial. They are mainly herbivores, but some are omnivores. The two groups of monkeys most probably were separated by the drifting apart of the continents of South America and Africa.

The apes are the chimpanzees, bonobos, and gorillas in Africa, and the orangutans of Asia. All are tropical and tree dwelling, except the gorillas, which are mainly terrestrial. All are opportunist omnivores and can, at times, hunt animals like monkeys. Gorillas live in packs lead by an elder male, or silverback, because of their grey hair. The popular view is that pack dominance is by threat and strength, but experience is the main factor. Younger and stronger pack members are prevented, by the rest of the pack, from challenging the silverback for leadership, until the youths have enough experience. The apes have very large brains and are capable of extremely complex behaviors, including tool use. Chimpanzees in the wild use straws to fish for termites in a termite mound (Goodall, 1986). The chimpanzees will break off multiple straws, go to the termite mound, put the straws down, and use one to fish for termites. When the first straw breaks, they take another and continue fishing for termites. Thus, the chimpanzees exhibit the complex psychological behavior of bringing a spare, which is anticipating the future. In captivity, they have demonstrated the ability to use logic to solve problems, even with insight as was first shown by Köhler (1925). They recognize themselves in mirrors. Washoe, a female chimpanzee, learned human sign language and could manipulate image containing cards. Chantele, an orangutan, learned human sign language and could give directions, when being driven for ice cream. Gorillas have learned human sign language. Probably, the most well-known of the captive ape projects is that of Penny Patterson (2003), who raised Koko, a female western lowland gorilla, from infancy for over 45 years. Koko learned to understand 2000 spoken English words and to use 1000 sign-language hand signs. Critics claim the English context is not correct. However, the capacity and communication is profound. Both Koko and a male companion, named Michael, created paintings. After occasions when acting aggressive, the apes explained they were angry. Thus, they recognize their internal emotions. Some captive apes can remember the events surrounding their capture. The apes are the closest relatives of humans. Rumbaugh and Washburn (2003) considered the intelligence of apes.

This Chapter has dealt with evolution and the increasing complexity of structure and of behavior that has resulted. Species are not superior or inferior, but rather, specialized into different niches.

References

Allen, R. D. (1962). "Amoeboid movement." *Scientific American*, February.

Applewhite, P. B. (1975). "Learning in bacteria, fungi, and plants." In *Invertebrate Learning*, Vol. 3, W. C. Corning, J. A. Dyal, and A. O. D. Willows, Eds., New York-London, Plenum Press, 179–186.

Arbit, J. (1957). "Diurnal cycles and learning in earthworms." *Science*, **126**, 654–655.

Arey, L. B. and Crozier, W. J. (1918). "The homing habits of the pulmonate mollusk Onchidium." *Proc. Nat. Acad. Sci.*, **4**, 319–321.

Arey, L. B. and Crozier, W. J. (1921). "On the natural history of Onchidium." *J. Exp. Zool.*, **32**, 443–502.

Arnold, C. (2015). "Free-ranging cats on Georgia's Jekyll Island tend to eat mostly amphibians and insects." *National Geographic*, September, 2015.

Au, W. W. L., A. N. Popper, and R. R. Fay, Eds. (2000). "Hearing by Whales and Dolphins," In Springer *Handbook of Auditory Research* series, Springer-Verlag, New York.

Baumann, F. (1929). "Experimente über den Geruchssinn und den Beuteerwerb der Viper (Vipera aspis L.)." *Zech. Vergl. Physiol.*, **10**, 36–119.

Best. J. B. (1963). "Protopsychology." *Scientific American*, **208**(2), 54–62.

Best, J. B. and I. Rubinstein. (1962). "Maze learning and associated behavior in planaria." *J. Comp. Physiol. Psych.*, **55**, 560–566.

Best, J. B., A. B. Goodman, and A. Pigon. (1969). "Fissioning in planarians: control by the brain." *Science*, **164**, 565–566.

Blum, H. F. (1968). *Time's Arrow and Evolution*. Princeton, NJ: Princeton Univ. Press.

Bonner, J. T. (1959). "Differentiation in social Amoeba." *Scientific American*, December, 152–162.

Bonner, J. T. (1963). "How slime molds communicate." *Scientific American*, August.

Bonner, J. T. (1967). *The cellular slime molds*. Princeton, NJ: Princeton University Press.

Bovard, J. F. (1918). "The transmission of nervous impulses in relation to locomotion in the earthworm." *Univ. Calif. Publ. Zool.*, **18**, 103–134.

Boycott, B. B. (1965). "Learning in the octopus." *Sci. Amer.*, **212**(3), 42–50.

Boycott, B. B. and R. W. Guillery. (1962). "Olfactory and Visual Learning in the Red-Eared Terrapin, Pseudemys Scripta Elegans (WIED.)." *Journal of Experimental Biology*, **39**, 567–577.

Brigham, R. M., E. K. V. Kalko, G. Jones, S. Parsons, H. J. G. A. Limpens. (2004). "Bat echolocation research tools, techniques and analysis." *Bat Conservation International*, Austin, Texas.

Budelmann, B.-U., (1970). "Die Arbeitsweise der Stratolithenorgane von Octopus vulgaris." *Z. Vergl. Physiol.*, **70**: 278–312.

Budelmann, B.-U., V. C. Barber, and S. West. (1973). "Scanning electron microscopical studies of the arrangements and numbers of hair cells in the statocysts of Octopus vulgaris, Sepia officinalis and Loligo vulgaris." *Brain Res.*, **56**, 25–41.

Budelmann, B.-U., and H. G. Wolff. (1973). "Gravity response from angular acceleration receptors in Octopus vulgarism." *Jour. Comp. Physiol.*, **85**, 283–290.

Burian, R. M. (1978). "A methodological critique of sociobiology." In A. L. Caplan (ed.) *The Sociobiology Debate*. New York: Harper and Row, 376–395.

Burian, R. M. (1981–82). "Human sociobiology and genetic determinism." *The Philosophical Forum*, **13**, 43–66.

Buytendijk, F. J. (1919). "Acquisition d'habitudes par des êtres unicellulaires." *Arch. Néerl. Physiol.*, **3**, 455–468.

Carr, A. (1965). "The navigation of the green turtle." *Scientific American*, **212**, 78–86.

Carr, A. (1967). "Adaptive aspects of the scheduled travel of Chelonia." *Proc. Ann. Biol. Colloq.*, Oregon State University, **27**, 35–36.

Casteel, D. (1911). "The discriminative ability of the painted turtle." *Journal of Animal Behavior*, 1, 1–28.

Clemens, W. A., Jr., J. D. Archibald, and L. J. Hickey. (1981). "Out with a whimper not a bang." *Paleobiology*, 7, 293–298.

Cohen, S. S. (1970). "Are/were mitochondria and chloroplasts microorganisms?" *Am. Scientist*, **58**, 281–289.

Copeland, M. (1930). "An apparent conditioned response in Nereis virens." *Jour. Comp. Psych.*, **10**, 339–354.

Corning, W. C. and E. R. John. (1961). "Effect of ribonuclease on retention of conditioned response in regenerating planaria." *Science*, **134**, 1363–1365.

Corning, W. C. and R. Von Burg. (1973). "Protozoa." In Invertebrate Learning, Vol. 1, W. C. Corning, J. A. Dyal, and A. O. D. Willows, Eds., New York-London, Plenum Press, 49–122.

Danisch, F. (1921). "Ueber Reizbiologie und Reizempfindlichkeit von Vorticella nebulifera." *Zech. allg. Physiol.*, **19**, 133–190.

Day, L. and M. Bentley. (1911). "A note on learning in Paramecium." *Jour. Anim. Behav.*, **1**, 67–73.

Dooling, R. J. (1982). "Auditory perception in birds." In *Acoustic Communication in Birds*, Vol. 1, D. E. Kroodsma and E. H. Miller, Eds., New York, Academic Press, 95–130.

Fischel, W. (1931). "Dressurversuche mit Schnecken." *Zsch. Vergl. Physiol.*, **15**, 50–70.

Fisher, H. (1898). "Quelques remarques sur les moeurs des Patells." *J. de Conchylol.*, **46**, 314–318.

Fisher, P-H. (1939). "Sur l'habitat et l'hygrophile des Succinées." *Jour. de Conchylol.*, **83**, 111–128.

Fisher, P-H. (1950). Vie et Moeurs des Mollusques. Paris, France.

Fox, S. W. et al. (1970). "Chemical origins of cells." *Chemical and Engineering News*, **48**: 80–94.

Fox, S. W. (1971). "Chemical origins of cells: 2." *Chemical and Engineering News*, **49**: 46–53.

French, J. W. V. (1940a). "Trial and error learning in paramecium." *Journal of Experimental Psychology*, **26**, 609–613.

French, J. W. V. (1940b). "Individual differences in paramecium." *Journal of Comp. Psychol.*, **30**, 451–456.

Friedländer, B. (1894). "Beitrage zur Physiologie des Zentralnervensystems und Bewegungs-mechanismus der Regenwürmer." *Pflüg. Arch. ges. Physiol.*, **58**, 168–207.

Garth, T. R. and M. P. Mitchell. (1926). "The learning curve of a land snail." *Jour. Comp. Psychol.*, **6**, 103–113.

Gelber, B. (1952). "Investigations of the behavior of Paramecium auratus: I. Modification of behavior after training with reinforcement." *Journal of Comparative and Physiological Psychology*, **45**, 58–65.

Goodall, J. M. (1986). *The chimpanzees of Gombe: patterns of behavior*. Boston: Belknap Press of Harvard University Press.

Gould, J. L. (1986). "The locale map of honey bees: Do insects have cognitive maps?" *Science*, **232**, 861–863.

Guhl, A. M. (1956). "The social order of chickens." *Scientific American*, February, 1–6.

Haldane, J. B. S. (1954). "The origin of life." In New Biology, London: Penguin Books, 12–27.

Hamilton, W. D. (1964). "The genetical theory of social behavior: I and II." *Journal of Theoretical Biology*, 7, 1–52.

Hecht, S. (1929). "The nature of photoreceptor process." In *Found. Exper. Psychol.*, Chapt. 5 (ed. by Murchinson). Worcester, MA: Clark Univ. Press.

Heck, L. (1920). "Über die Bildung einer Assoziation beim Regenwurm auf Grund von Dressurversuchen." *Lotos Naturwiss. Zech.*, **68**, 168–189.

Hewatt, W. G. (1940). "Observations on the homing limpet, Acmaea scabra Gould." *Am. Midl. Naturalist*, **24**, 205–208.

Hochner, B. and D. L. Glanzman. (2016). "Evolution of highly diverse forms of behavior in molluscs." *Current Biology*, **26**(20), R965–R971.

Hölldobler, B. and E. O. Wilson. (1990). *The Ants*. Cambridge, MA: The Belknap Press of Harvard University Press.

Hölldobler, B. and E. O. Wilson. (2011). *The leafcutter ants, civilization by instinct*. W. W. Norton & Company, Inc., New York, NY.

Horridge, G. A. and B. MacKay. (1962). "Naked axons and symmetrical synapses in coelenterates." *Quart. J. Microscop. Sci.,* **103**, 531–541.

Jacobson, A. L. (1963). "Learning in flatworms and annelids." *Psychological Bulletin*, **60**, 74–94.

Jacobson, A. L. (1965). "Learning in planarians: Current status." *Animal Behaviour, Suppl.,* **1**, 76–80.

Jennings, H. S. (1901). "Studies on reactions to stimuli in unicellular organisms. IX. On the behavior of fixed infusoria (Stentor and Vorticella) with special reference to modifiability of protozoan reactions." *Am. J. Physiol.,* **8**, 23–60.

Jennings, H. S. (1906). *Behavior of the lower organisms*. (1923ed.) Columbia University Press, New York, 188–216.

Jensen, D. D. (1957a). "Experiments on "learning" in paramecia." *Science*, **125**, 191–192.

Jensen, D. D. (1957b). "More on "learning" in paramecia." *Science*, **126**, 1341–1342.

Jha, R. K. and G. O. Mackie. (1967). "The recognition, distribution and ultrastructure of hydrozoan nerve elements." *J. Morp.,* **123**, 43–61.

Kettlewell, H. B. D. (1959). "Darwin's missing evidence." *Scientific American*, **200**(3), 48–53.

Kettlewell, H. B. D. (1965). "Insect survival and selection for pattern." *Science*, **148**, 1290–1296.

Köhler, W. (1925). *The Mentality of Apes*. Harcourt Brace, New York.

Konishi, M. and E. I. Knudsen. (1979). "The oilbird: hearing and echolocation." *Science*, **204**: 425–427.

Kreithen, M. L. and T. Eisner. (1978). "Ultraviolet light detection by the homing pigeon." *Nature*, London, **272**, 347–348.

Kreithen, M. L. and W. T. Keeton. (1974). "Detection of changes in atmospheric pressure by the homing pigeon, Columba livia." *J. Comp. Physiol.* **89**, 73–82.

Kreithen, M. L. and D. B. Quine. (1979). "Infrasound detection by the homing pigeon: A behavioral audiogram." *J. Comp. Physiol. A.,* **129**, 1–4.

Kuenzer, P. P. (1958). "Verhaltenphysiologische Untersuchungen über das Zucken des Regenwurms." *Zeit. Tierpsychol.,* **15**, 31–49.

Lindauer, M. (1961). *Communication among social bees*. Cambridge, MA: Harvard University Press.

Lissman, H. W. (1963). "Electric location by fishes." *Scientific American*, March.

Loomis, W. F. (1967). "Skin pigment regulation of Vitamin-D biosynthesis in man." *Science*, **157**, 501–506.

Losina-Losinsky, L. K. (1931). "Zur Ernåhrungsphysiologie der Infusorien Untersuchungen über die Nahrungsauswahl und Vermehrung bei Paramecium caudatum." *Arch. Protistenk*, **74**, 18–120.

Lowontin, R. C. (1977). "Sociobiology- a caricature of Darwinism." In P. Asquith and F. Suppe (Eds.), *The Philosophy of Science Association*, **2**, 22–31.

Maier, N. R. F. and T. C. Schneirla. (1935). *Principles of Animal Psychology*. New York: McGraw-Hill, reprinted by Dover, N. Y. (1963).

Margulis, L. (1971). "The origin of plant and animal cells." *American Scientist*, **59**, 230–235.

Mast, S. O. and L. C. Pusch. (1924). "Modification of response in Amoeba." *Biol. Bull.*, **46**, 55–60.

Mason, J. R., D. A. Stevens, and M. D. Rabin. (1980). "Instrumentally conditioned avoidance by tiger salamanders (Amblystoma tigrinum) to reagent grade odorants." *Chem. Senses*, **5**: 99–103.

Mazzeo, R. (1953). "Homing of the Manx Shearwater." *Auk*, **70**, 200–201.

McConnell, J. V. (1966). "Comparative physiology: Learning in invertebrates." *Annual Review of Physiology* (Palo Alto, Calif.: Annual Reviews, Inc.), **28**, 107–136.

McConnell, J. V., A. L. Jacobson, and B. M. Humphries. (1961). "The effects of ingestion of conditioned planaria on the response level of naive planaria: a pilot study (or: "you are what you eat?")." *Worm Runner's Digest*, **3**, 41–47.

McConnell, J. V., A. L. Jacobson, and D. P. Kimble. (1959). "The effects of regeneration upon retention of a conditioned response in the planarian." *Journal of Comparative Physiological Psychology*, **52**, 1–5.

McConnell, J. V. and D. H. Malin. (1973). "Recent Experiments in Memory Transfer." In: Zippel H. P. (Eds.) *Memory and Transfer of Information*, Plenum Press, New York.

Metalnikow, S. (1912). "Contributions `a l'étude de la digestion intracellulaire chez les protozoaires." *Arch. de Zool.*, **49**, 373–498.

Miller, S. L. (1953). "A Production of Amino Acids Under Possible Primitive Earth Conditions." *Science*, **117**(3046): 528–529.

Miller, S. L. and H. C. Urey. (1959). "Organic Compound Synthesis on the Primitive Earth." *Science*, **130**(3370): 245–251.

Mirsky, S. (2008). "Shakespeare to blame for introduction of European starlings to US." *Scientific American*, June.

Morton, J. E. (1962). "Habit and orientation in the small commensal bivalve mollusk, Montacuta ferruginora." *Animal Behav.*, **10**, 126–13

Oh, D.-C., M. Poulsen, C. R. Currie, and J. Clardy. (2009). "Dentigerumycin: a bacterial mediator of ant-fungus symbiosis." *Nature Chemical Biology*, **5**, 391–393.

Olmsted, J. (1922). "The role of the nervous system in the locomotion of certain marine polyclads." *Jour. Exper. Zool.*, **36**, 57–66.

Oparin, A. I. (1968). *Genesis and Evolutionary Development of Life*. Translated by Eleanor Maass. New York: Academic Press.

Patterson, F. G. P. (2003). "Communication Studies" in the *Great Ape Project Census: Recognition for the Uncounted*. Great Ape Project Books, 219–232.

Pelseneer, P. (1935). "Essay on zoological ethology based on the study on Mollusca [Essai d'éthologie zoologique d'après étude des mollusques.]." *Académie royal de Belgique*, Bruxelles. Publications de la Foundation Agathon de Potter, No. 1, 662 pp.

Pierce, N. E. (1985). "Lycaenid butterflies and ants: selection for nitrogen-fixing and other protein-rich food plants." *American Naturalist*, **125**(6) 888–895.

Pieron, H. (1909). "Contribution a la biologie de la Patelle et de la Calyptreé. I. L'éthologie, les phénomènes sensoriels." *Bull. Sci. Fr. Belg.*, **43**, 183–202. "II. Le sens du retour et la mémoire topographique." *Arch. Zool. Exp. Gén.* (5) Notes et Revues, 18–29.

Ratner, S. C., and K. R. Miller. (1959a). "Classical conditioning in earthworms, Lumbricus terestris." *J. Comp. Physiol. Psychol.*, **52**, 102–105.

Ratner, S. C., and K. R. Miller. (1959b). "Effects of spacing of training and ganglion removal on conditioning in earthworms." *J. Comp. physiol. Psychol.*, **52**, 667–672.

Rau, P. and N. Rau. (1931). "Additional experiments on the homing of carpenter- and mining-bees." *Jour. Comp. Psychol.*, **12**, 257–261.

Ritter, K. S., B. A. Weiss, A. L. Norrbom, and W. R. Nes. (1982). "Identification of $\Delta^{5,7}$-24-methylene and methyl sterols in the brain and whole body of Atta ciphalotes isthmicola." *Comparative Biochemistry and Physiology B*, **71**, 345–349.

Robertson, D. R. (1972). "Social control of sex reversal in a coral-reef fish." *Science*, **177**, 4053, 1007–1009.

Robinson, J. S. (1953). "Stimulus substitution and response learning in the earthworm." *J. Comp. Physiol. Psychol.*, **46**, 262–266.

Ross, D. M. (1965). "Behavior of sessile coelenterates in relation to some conditioning experiments." *Animal Behaviour, Suppl.*, **1**, 43–53.

Rumbaugh, D. M. and D. A. Washburn. (2003). *The intelligence of apes and other rational beings.* New Haven, CT: Yale University Press.

Rushforth, N. B. (1965a). "Behavioral studies of the coelenterate Hydra pirsrdi Brien." *Animal Beahviour, Suppl.*, **1**, 30–42.

Rushforth, N. B. (1965b). "Inhibition of contraction responses of Hydra." *Am. Zoologist*, **5**, 505–513.

Rushforth, N. B. (1967). "Chemical and physical factors affecting behavior in Hydra: Interactions among factors affecting behavior in Hydra." In Corning, W. C., and S. C. Ratner (eds.), *Chemistry of Learning*, Plenum Press, New York.

Rushforth, N. B. (1970). "Electrophysiological correlate of habituation in Hydra." *Am. Zoologist.* **10**, 505.

Rushforth, N. B. (1973). "Behavioral Modifications in Coelenterates." In Invertebrate Learning, Vol. 1, W. C. Corning, J. A. Dyal, and A. O. D. Willows, Eds., New York-London, Plenum Press, 123–169.

Rushforth, N. B. and D. S. Burke. (1971). "Behavioral and electrophysiological studies of Hydra. II. Pacemaker activity of isolated tentacles." *Biol. Bull.*, **140**, 502–519.

Rushforth, N. B., A. L. Burnett and R. Maynard. (1963). "Behavior in Hydra. Contraction responses of Hydra pirardi to mechanical and light stimuli." *Science*, **139**, 760–761.

Sanders, G. D. (1975). "The Cephalopods." In *Invertebrate Learning*, Vol. 3, W. C. Corning, J. A. Dyal, and A. O. D. Willows, Eds., New York- London, Plenum Press, 1–101.

Sauer, E. G. F. (1971). "Celestial rotation and stellar orientation in migrating warblers." *Science*, **173**, 459–460.

Schneider, D. (1974). "The sex-attractant receptor of moths." *Scientific American*, July.

Schneirla, T. C. (1934). "The process and mechanism of ant learning." *J. Comp. Psychol.*, **17**, 303–328.

Schneirla, T. C. and G. Piel. (1948). "The army ant." *Scientific American*, **178**, 6, June: 16–23.

Schneirla, T. C. (1959). "An evolutionary and developmental theory of biphasic processes underlying approach and withdrawal." *Nebraska symposium on motivation.* M. R. Jones (ed.) Lincoln: Univ. of Nebraska Press, 1–42.

Schöne, H. and B.-U. Budelmann. (1970). "Function of the gravity receptor of Octopus vulgarism." *Nature*, **226**, 864–865.

Schopf, J. W., (Ed.), (1983). *Earth's Earliest Biospheres, Its Origin and Evolution*. Princeton, NJ: Princeton University Press.

Simmons, J. A. and R. A. Stein. (1980). "Acoustic imaging in bat sonar: echolocation signals and the evolution of echolocation." *J. Comp. Physiol. A.*, **135**(1): 61–84.

Skinner, B. F. (1956). "A case history in scientific method." *Amer. Psych.*, **11**, 221–233.

Smith, S. (1908). "The limits of educability in Paramecium." *Jour. Comp. Neurol. Psychol.*, **18**, 499–510.

Stasko, A. B. and C. M. Sullivan. (1971). "Responses of planarians to light: an examination of kino-kinesis." *Animal Behavior Monographs*, **4**, part 2.

Thompson, E. (1917). "An analysis of the learning process in the snail, Physa gyrina Say." *Behav. Monogr.*, **3**, 5–39.

Thompson, R. T. and J. V. McConnell. (1955). "Classical conditioning in the planarian, Dugesia dorotocephala." *J. Comp. Physiol. Psych.*, **48**, 65–68.

Thorpe, W. H. (1956). *Learning and Instinct in Animals*. Methuen, London.

Thorpe, W. H. (1963). *Learning and Instinct in Animals*. (2nd ed.) Cambridge: Harvard University Press.

Todd, K. (2001). *Tinkering with Eden: a natural history of exotics in America*, W. W. Norton & Co., New York.

Trivers, R. L. (1972). "*Parental investment and sexual selection*." In B. Campbell (Ed.), Sexual selection and the descent of man. Chicago: Aldine.

Underwood, B. J. (1966). *Experimental Psychology*. Appleton-Century-Crofts, New York.

von Frisch, K. (1971). *Bees: Their vision, chemical senses and language*. Ithaca, NY: Cornell University Press.

Weber, N. A. (1972). *Gardening ants: the attines*. (Memoirs of the American Philosophical Society, vol. 92). American Philosophical Society, Philadelphia.

Weiss, B. A. (1966). "Hearing in the goldfish (Carassius auratus)." *Journal of Auditory Research*, **6**, 321–335.

Weiss, B. A. (1969). "Lateral-line sensitivity in the goldfish, (Carassius auratus)." *Journal of Auditory Research*, **9**, 71–75.

Weiss, B. A. (1990). "Longitudinal observation of attine (Atta cephalotes isthmicola) ants." *Zoo Biology*, December **6**, 421–429.

Weiss, B. A. and W. F. Strother. (1965). "Hearing in the green treefrog (Hyla cinerea)." *Journal of Auditory Research*, **5**, 297–305.

Weiss, B. A. and T. C. Schneirla. (1967). "Inter-situational transfer in the ant (Formica schaufussi) as tested in a two-phase, single choice-point maze," *Behaviour*, **27**, 269–279.

Weiss, B. A., G. M. Hartig, and W. F. Strother. (1969). "Hearing in the bullhead catfish (Ictalurus nebulous)." *Proceedings of the National Academy of Science*, **64**, 552–556.

Weiss, B. A. and J. L. Martini. (1970). "Lateral-line sensitivity in the blind cavefish (Anoptichthys jordani)." *Journal of Comparative and Physiological Psychology*, **71**, 34–37.

Weiss, B. A., B. H. Stuart, and W. F. Strother. (1973). "Auditory sensitivity in the tadpole (Rana catesbeiana)." *Journal of Herpetology*, **7**, 211–214.

Wells, M. J. (1962a). *Brain and Behaviour in Cephalopods*. Stanford University Press, Stanford, Calif.

Wells, M. J. (1962b). "Early learning in Sepia." *Sym. Zool. Soc. Long.*, **8**, 149–169.

Wenner, A. M. (1971). *The bee language controversy*. Boulder, CO. Educational Programs Improvement.

Werner, Y. L. (2016). *Reptile life in the land of Israel with comments on adjacent regions*. Brahm, Frankfurt am Main, Germany.

Westfall, J. A. (1969). "Ultrastructure of synapses in a primitive coelenterate." *J. Ultrastruct. Res.*, **32**, 237–246.

Westfall, J. A. (1970). "Synapses in a sea anemone, Metridium (Anthozoa)." *7th Congr. Internat. Microscop. Electron* Grenoble, 717–718.

Westfall, J. A., S. Yamataka, and P. D. Enos. (1970). "Ultrastructure of synapses in Hydra." *J. Cell. Biol.*, **47**, 266.

Westfall, J. A., S. Yamataka, and P. D. Enos. (1971). "Ultrastructural evidence of polarized synapses in the nerve net of Hydra." *J. Cell Biol.*, **51**, 318–323.

Wever, E. G. (1949). *Theory of hearing*. New York: John Wiley.

Wever, E. G. (1978). *The Reptile Ear*. Princeton, NJ, Princeton University Press.

Whittaker, R. H. (1969). "New concepts of kingdoms of organisms." *Science*, **163**, 150–160.

Willows, A. O. D. (1973). "Learning in Gastropos Mollusks." In *Invertebrate Learning*, Vol. 2, W. C. Corning, J. A. Dyal, and A. O. D. Willows, Eds., New York-London, Plenum Press, 187–273.

Wilson, E. O. (1971). *Insect Societies*. Cambridge, MA: The Belknap Press of Harvard University Press.

Wiltschko, R., and Wiltschko, W. (1995). *Magnetic orientation in animals*. New York, Springer-Verlag.

Wolff, H. G. (1970). "Statocystenfunktion bei einigen Landpulmonaten (Gastropoda)." *Z. Vergl. Physiol.*, **69**, 326–366.

Yerkes, R. M. (1901). "The formation of habits in the turtle." *Pop. Sci. Mo.*, **58**, 519–525.

Yerkes, R. M. (1912). "The intelligence of earthworms." *Jour. Anim. Behav.*, **2**, 332–352.

Young, J. Z. (1972). *The Anatomy of the Nervous System of Octopus vulgaris*. New York: Oxford University Press.

CHAPTER 3
Nature in Humans

Around 55 million years ago, small primates and rodents were living on the ground in the wake of the great dinosaur extinction. The rodents mainly remained on the ground, but the primates took to the trees. The ground and tree environments are very different. Ground life is heavily subject to predation and, thus, necessarily nocturnal. Nocturnal living means selection for sensitive night vision with no color vision, which would detract from low-light sensitivity. Eyes need to face sideways to provide the maximum range of vision to prevent surprise.

Tree life means daytime living to see the next branch well. Jumping from branch to branch requires accurate vision, with color perception, and front-facing eyes to judge the position and distance of the next branch. Not having a wide range of vision does not risk predation, because predators must climb the tree. Shaking the tree by climbing provides warning. The primates would sleep in the trees at night. Only those rodents that took to the trees, like some squirrels, would share the adaptations of the primates.

Thus, the primates would evolve to be well suited to tree life. During the warm climate of the time, primates evolved into many niches in the forest canopy. They also spread across Africa and into Europe and Asia. South America, being closer to Africa, was also inhabited by primates. The evolution of primates into tree life lasted for many millions of years, with adaptations, like a fist grip for holding onto branches.

Then, the climate cooled. Primate life became more restricted to equatorial areas. In East Africa, in the geological fault called the Rift Valley, the land became elevated. The trees of the forest became sparse. The result was woodland savanna. Those primates least suited for tree-life were forced onto the ground. They adapted by grabbing fruit from a tree and running upright with an arm load into another tree. Thus, began the development of running and an upright stance, which also, raised eye level for quicker detection of predators. On the ground, a lone primate will easily be captured by a predator. Therefore, primates can only survive in groups. The emotional calls of tree-living primates became refined for more accurate communication for group defense and hunting on the ground.

People consider language a human intellectual triumph. However, language is hardly intellectual, having evolved for communication in group defense and hunting. Language appearance in the individual is correlated with motor development (Lenneberg, 1967, 1969). At a particular age, a child has certain motor skills and corresponding language ability. That developmental sequence is the same in all cultures and every language. If a child arrests at a certain motor level, then language development remains at the corresponding ability. Specific brain damage affects language. If language were an intellectual skill, then there would be some culture without language development. All cultures have spoken language. Thus, spoken language evolved from early human pack defense and hunting. Written language is different and was invented recently. People first tried to write visually with pictures. That attempt eventually became sophisticated in forms like Chinese. But picture writing requires so many characters that only scholars in that language can become proficient. The majority of people manage with a smaller knowledge of characters. Written language was revolutionized by the invention in Semitic writing of the alphabet. The alphabet, with a small number of letters, permitted portraying the sound of speech, rather than pictures. Alphabetic writing could be mastered by essentially everyone.

When early primates on the ground detected a predator they issued an alarm call. The pack would respond by throwing sticks and stones at the attacker. The volley of sticks and stones pummeling the predator caused hesitation, allowing the pack to escape into the trees. Kortlandt (1962) reports chimpanzees in the open will not only throw objects at predators, but will attack with stick clubs.

The poorer climbing ability of the those early primates, caused them to be crowded onto the ground, but provided arms with good throwing ability, especially the side-arm style of throwing. Throwing objects led to development of better and better throwing instruments. Grip enhancement, stone chipping, and stick sharpening became refined. Rocks were chipped into sharp-edged tools and branches were sharpened into spears. The construction of tools, starting in primates with smaller brains, required increasing dexterity and skill with more elaborate sensory, motor, and brain capacity. Thus, tool use caused the evolution of the brain. The evolution of the primates into human beings had begun. Eventually, the biggest enlargement in the human brain would be threefold: the incoming sensory display, the outgoing motor response area, and the largest growth was in the association region that connects sensory input with motor output. Osiurak (2017) reviewed and discussed the cognitive evolution necessary for tool use.

The use of tools at the ends of arms relieved the selection favoring the protruding jaw and large teeth that were previously required as weapons. A smaller jaw needed less of a protruding brow ridge, where muscles could be attached to operate the jaw. Releasing the face from functioning as a weapon would protect the head. The evolution of the large brain required a tradeoff in head size. The face became flattened and reduced, especially, the protruding large jaw and the size and number of teeth. That evolution continues in the present with people being born with impacted or totally absent rear molars, referred to as "wisdom" teeth.

Why would a tradeoff be required? The behavior of the early hominin pack was hunting by running after prey and killing it. Without refrigeration, the kill would last a day or so, requiring another hunt. A large pack would need to kill a sizable animal. The whole pack would have to participate in the hunt. More vulnerable individuals, like females and young, could not be left behind because of predators. The female pelvis can not be too much larger than that of the male, or the running of the pack would be slowed. Since, the head of the infant at birth must pass through the female pelvis, and expansion is limited by running requirements, the evolution of the large brain came at the expense of the protruding face and jaw.

In modern terms, a horse can run a mile in less than three minutes, while humans take much longer. But horses evolved for sprints to avoid the charge of a predator. For a horse to run a marathon distance would take most of a day. Humans, by contrast, can run a marathon in hours. Thus, the early hominin pack could run down a large animal by continued pursuit. A few hours of running to capture a couple of days of food would be efficient behavior.

The first animal to associate with hominins was the wolf. A pack of wolves (dogs) hunts the same way as the early hominin pack did. They would become excited by the hunt. They would run ahead of the hominin pack and slow down the prey for the hominins to catch up and kill it with their weapons. Sharing the food would continue the hunting association. Favoring the friendly wolves with shared food would result in the creation of the dog breed. Apparently, variants in the GTF21 and GTF21RD1 genes, implicated in the Williams-Beuren syndrome characterized by hyper social behavior in humans, was also being selected in favoring friendly wolves (von Holdt, et al., 2017). The experiment by Dimitri K. Belyaev (Director of the Institute of Cytology and Genetics of the Russian Academy of Science, 1959–1985) in breeding successive generations of foxes for friendliness produced a dog-like fox, in behavior, and even in the spotted or piebald coloration and tail wagging (Goldman, 2010, Ratliff, 2011). The fox experiment demonstrates how dogs were bred from wolves.

Comparing modern humans with apes shows humans have a higher brain weight, a larger head size relative to the body, a straighter angle of the head with the trunk, a more upright posture, slower growth, retarded closure of the sutures of the skull bones, smaller teeth, a flatter face, and general body hairlessness. These characteristics of adult humans match more closely to the structure and features of baby apes. Thus, the sequence of evolution, which began with the early hominid packs, was accomplished through the process of neoteny. In neoteny, the eventual form is achieved by earlier and earlier sexual maturity, until the present version is a mature human with a baby ape body. As adult humans age, they look more and more ape-like.

Because a lone primate would quickly be attacked by a predator, the group was required for survival. The hominin pack became the successful unit of behavior, enabling both organized hunting and defense. The glue for the pack was positive emotional identification with the pack and negative emotional response against others, in an "us against them" mentality. Structural features like sexual receptivity, separated from reproductive cycles, would also keep the pack together.

Although sparse in population, the packs spread across Africa, evolving new forms. Eventually, they migrated into Europe and Asia and evolved into Homo heidelbergensis. Homo heidelbergensis, thereafter, evolved into Homo neanderthalensis in Europe and West Asia, Homo denisovans in East Asia, and Homo sapiens (Cro-magnons) in Africa.

Neanderthals thrived in sub-glacial Europe from 350,000 to 28,000 years ago. Being in a cold region, they developed robust bodies to conserve heat. With the less intense sunlight of the northern latitudes, they evolved less pigmentation in skin, eyes, and hair. They survived in Europe and West Asia for over 300,000 years.

Around 40,000 years ago, Homo sapiens, which likely existed in Africa from 300,000 years ago, crossed into West Asia and Western Europe at both ends of the Mediterranean Sea. They encountered H. neanderthalensis in areas like the Levant and co-existed for awhile. But, within about 10,000 years Homo sapiens displaced H. neanderthalensis completely. The encounter of Homo sapiens with H. neanderthals resulted in possible killing, probable out competing of H. neanderthals by faster Homo sapiens, who had better tools, especially throwing spears, and also likely swamping of H. neanderthalensis by larger populations of Homo sapiens. In addition, there was definite interbreeding. Modern DNA analysis shows as much as 4% of H. neanderthal genes in all non-African people, including characteristics like fair skin, red hair, blue eyes, and barrel chests. East Asians also have as much as 6% H. denisovan genes.

By 30,000 years ago, H. Sapiens had spread across Europe and West Asia, into Southeast Asia and Australia. East Asia was populated by 25,000 years ago.

There were no indigenous, or native people, in North America. In *Across Atlantic Ice*, Stanford and Bradley (2012) describe the first people reaching

North America as coming from what now is Southern France. They traveled across the edge of the Atlantic Ice Age Shelf by hunting and fishing, perhaps using boats, during the Last Glacial Maximum. The distance across the Atlantic would have been less than now, because of the large amount of water locked into ice, because the continents had not drifted as far apart, and because the distance was from between the exposed continental shelves of Europe and America. They reached the area of the Chesapeake Bay, south of the North American continental ice sheet 15,000 years ago. With them came their Solutrean stone-point technology. As the climate warmed, settlements were established in places like Meadowcroft Rockshelter in Pennsylvania, Cactus Hill in Virginia, Miles Point and Oyster Cove in Maryland and Cinmar on the exposed outer continental shelf near the Chesapeake Bay. Cold and dry weather, called the Younger Dryas, returned about 13,000 years ago. The people of the Eastern settlements were forced South and West. They encountered the "Clovis" people who were arriving over the Bering land bridge from Siberia to Alaska as the Last Glacial Maximum ice retreated. The Clovis people had moved south with their Siberian bone tools. With the encounter, the Clovis people acquired the stone Solutrean tool culture and, thereafter, spread into Central and South America. A European mitochondrial genetic marker exists in pre-Columbian human remains and in some modern American Indians, but not in Asian populations.

The retreat of the ice cut the cross Atlantic route but further opened the Asian migration. Settlement from across the Atlantic would not occur again for millennia, when the voyages of the Vikings and Columbus reopened the cross Atlantic route. Columbus is credited with the discovery of the Americas, even though he believed he had found a new route to the East Indies, because his achievements were generally known. Da Vinci, for example, had interesting scientific ideas, but hid most of them, making the ideas unknown until they were long surpassed. Thus, he is known as an artist, but not as a scientist.

Subsequently, people from Southeast Asia reached the Pacific Islands. Thus, the initial inhabitation of the World by humans was complete by 10,000 years ago. Human society included improved tools and weapons, use of fire, domesticated animals, like dogs, sheep, cattle, horses, among others, and living in settlements with agriculture. Until the industrial revolution, all work was accomplished by humans, including slaves in many societies, by using animals, and by some tool usage. After the industrial revolution slavery has almost vanished in modern societies.

Note: The persistent reports of a large ape in east Asia and North America are intriguing to scientists, because there was a giant ape. The giant ape, called Giantopithecus, went extinct about 100,000 years ago, possibly because of competition from the spreading Homo populations. Some estimates put Giantopithecus at a height of 3m (9.8 ft) and a weight of up to 540 kg (1190 lbs). Giantopithecus appears to have been a relative of orangutan. Fossil teeth indicate that Giantopithecus was plant eating, like the giant panda. If Giantopithecus survived in the remote areas where the reports are originating on both sides of the former Siberian-Alaskan land bridge, that would provide an explanation for the reports. However, so far, there has been no scientific substance to the sightings, photography, foot prints, or relics that have been reported.

References

Goldman, J. G. (2010). "Man's new best friend? A forgotten Russian experiment in fox domestication." *Scientific American*, September.

Kortlandt, A. (1962). "Chimpanees in the wild." *Scientific American*, May.

Lenneberg, E. H. (1967). *Biological Foundations of Language*. New York: Wiley.

Lenneberg, E. H. (1969). "On explaining language." *Science*, **165**, 635–643.

Osiurak, F. (2017). "Cognitive Paleoanthropology and Technology: Toward a Parsimonious Theory (PATH)." *Review of General Psychology*, **21**, 4, 292–307.

Ratliff, E. (2011). "Taming the Wild." *National Geographic*, March, 40–59.

Stanford, D. J. and B. A. Bradley. (2012). *Across Atlantic Ice, Berkeley*, Calif., University of California Press.

von Holdt, B. M., E. Shuldiner, I. J. Koch, R. Y. Kartzinel, and A. Hoga. (2017). "Structural variants in genes associated with human Williams-Beuren syndrome underlie stereotypical hyper sociability in domestic dogs." *Science Advances*, **3**, 7, July 19.

CHAPTER 4

Humans in Nature

Humans made their population expansion across the globe, during, and at the end, of the last Ice Age. However, the climate of the Earth has never been stable. It has varied between an ice covered planet to ice-free periods. For example, the age of the dinosaurs was much warmer than now, with ten times higher levels of atmospheric carbon dioxide and no polar ice. Over the last 350,000 years there have been four major cycles of increased atmospheric warming, followed by Ice Ages between the warm periods.

Three major factors determine the climatic conditions of the Earth. The first and foremost is the energy output of the sun, called insolation, which is variable. The second is the orbit of the Earth, which currently is almost circular, but it can be elliptical. The third is the tilt of the axis of rotation of the Earth, with respect to the plane of the orbit. The tilt, now near 23.5°, can vary from 21.8° to 24.4°. The further the orbit is from round and the larger the tilt of the axis of spin, the more extreme are the seasonal changes from winter to summer. An ice age builds when more snow falls and ice forms in the winter than melts in the summer.

The composition of the atmosphere can also affect the climate. The sun heats the surface of the Earth. Gases like methane, carbon dioxide, sulfur dioxide, and water vapor reflect back escaping surface heat, raising the temperature. Thus, these gases are called "greenhouse gases." Major sources of these gases are volcanic eruptions, mantle venting from faults in the crust, forest fires, and biological activity. Greenhouse gases from biological activity includes plant releases and absorption, human and animal production, as well as, plant and animal decay. The photosynthetic organisms on land and in the ocean consume gases like carbon dioxide, but in less quantities than are being produced. The colder regions of the oceans are a storehouse of these gases. Ocean currents can sequester quantities of greenhouse gases, even to the point of masking the amount produced. When the oceans warm, the stored gases are released, much like carbonated soda releases carbon dioxide as it warms. The sun has also warmed other planets, including Jupiter and Mars, recently without the contribution of greenhouse gases.

Human greenhouse gas production has been decreasing, but, is overwhelmed by the other factors, for example, the currently extensive world-wide forest fires, which are largely human caused. Conservation remains valuable, even though global temperature has been stable for decades (Kennedy, 2013). Global warming with CO_2 will improve plant growth and habitation in Alaska, Canada, Northern Europe, and Siberia. However, since 1880, there has only been a 1 °C (1.7 °F) increase (NOAA, 2017). Human records, span only a few centuries, and are not extensive compared to millions of geological years. Prediction is uncertain and warming estimates have been incorrect (Happer, 2012). However, with the periodic cycles between warm and cold, another "mini-ice age" is possible. Lindzen (2017) reviewed all the arguments for global warming, including the ideas of consensus, record heat, extreme weather, melting sea ice, endangered polar bears, and ocean acidification, and finds them lacking any scientific evidence, but still being employed for swaying public policy. Warming has ceased with current global temperature stable since 1998. That may be a long pause or a peak.

In past centuries, the Earth's climate has been very variable. In the Middle Ages, the climate was warmer than now. The Vikings were a sea power, invading Europe. They also settled in, at that time, a warmer Iceland, Greenland, and Eastern Canada. Around 1300 CE the climate turned much colder into the "Mini-Ice Age." The cold finally finished the Vikings, who abandoned their Canadian settlements and lost their high military status because northern rivers and ports froze. In 1776 CE, General George Washington launched a Christmas attack from Pennsylvania against the Hessians at Trenton, NJ, and the British in Princeton, NJ. He anticipated the Delaware River, which separated the opposing forces, and usually froze, would thus provide his adversaries with access to his own Army. Since around 1850 CE, before the industrial revolution, the climate has been warming. Thus, currently, the Delaware River rarely ever freezes.

Regardless of climate, humans have continued to move. The Mongols invaded Europe from Asia. Europeans moved into the Americas, bringing about 6 million African slaves, 90% to the Caribbean and South America. These and other movements all caused fatalities from warfare and disease, but altered the human population distribution. Diseases like smallpox and measles spread to the Americas, while, syphilis spread to Europe and beyond.

The effect humans have on the environment can also be extensive. When early people reached North America they encountered many animals that were, thereafter, hunted to extinction. Those animals included mammoths, ground sloths, camels (which migrated into Asia, where they exist as the familiar camels, and into South America, where they exist as llamas and their relatives), and many others. Krantz (1970) explained how humans cause extinction. We compete for a particular food supply. Killing mammoths would lead to the loss of mammoth predators like saber-tooth cats. Humans hunt by killing large males and leaving females and young. That leads to a population boom. Natural predation, where young and infirm are culled, stabilizes the population size. The population boom, caused by human hunting, in one species eliminates others. For example, a bison boom eliminates horses, cattle, and camels. Mammoth hunting did not lead to a boom in population, because of the long gestation period. Also, agricultural practices, providing food for human settlements, causes hunting for trophies of rare animals to trade for crops. Thus, the last few of a species are eliminated. Finally, the presence of a human population can prevent animal access to food and water. We also cause plants and animals to become invasive into other territories. In the exchange after Columbus, tomatoes, potatoes, corn, pumpkins, chocolate, coffee, tobacco, some hallucinogens, turkeys and more went to Europe. From Europe came many crops and animals like pigs, horses, cattle, rats, and more, with people. People cut most of the forrest and removed many plants for crop farming in the Americas.

Human success has been due to the pack organization, which allows for common defense, hunting, and food production. The pack retains its structure through positive emotional identification by individuals with their own pack, and negative emotional response against others. These emotional responses are human instincts and the cause of prejudices. People have long been interested in what is instinct and what is learning, or what is nature and what is nurture. Herodotus, the ancient Greek historian, noted that centuries ago Psammetichus, an Egyptian Pharaoh, ordered two children reared without ever hearing anyone speak. When the children were older, they spoke Phrygian, proving that language was innate and that Phrygian was the oldest language (Ravolinson, 1952). Apparently, the Pharaoh did not imagine that Phrygian servants had defied his orders and helped the children.

At the end of the section about arthropods in Chapter 2, the concepts of instinct and learning were analyzed as the opposite ends of a continuous dimension of modifiability. Current ideas conclude that humans do not have instincts, but we do. Emotional responses are human instincts. Emotional identification with a pack is a strong human instinct, because, in human history, a lone primate would be killed by predators. Humans could only exist in packs to successfully defend themselves and hunt.

Humans now have layers of pack identities. Some of the layers include, for example, their sports teams, their cities, their ethnic groups, their religions, their political loyalty, and their nations. Those within a group can be positive or critical, but negative attitudes from outsiders will not be tolerated. Even after a competitive sports event, like a soccer match, riots can result in fatalities. People strive to belong to their packs. Expressions of identity include adopting group trends in clothes, music, political slogans, and attitudes. People accept and follow the ideology of their group, becoming irrational, emotional trendies with little thought. Politicians, who can invoke a sense of pack identity in their constituents, will be elected. Merchandise that arouses positive emotional response will sell.

Pack doctrine must be followed or ostracism could occur. Group identity is not rational, but requires emotional loyalty (Hoffer, 1951). Belonging to a pack reduces stressful decisions by the individual, who yields intellectually to the pack doctrine (Fromm, 1941). All packs are not equally valid. Some do well, some provide poorly, and some have vanished. The result is hostility among rival groups. Hostility can be expressed in behavior leading from mild competition up to outright warfare. Past pack hostilities have resulted in major wars with many millions of deaths. Pack arsenals now include nuclear weapons. For the last half of the 20th century, the United States and the USSR were able to avoid nuclear war. However, proliferation of nuclear weapons among many nations, with more seeking to have the weapons, makes nuclear war very likely.

A major pack achievement has been increasing human populations. The total number of people on Earth reached one billion about 1850 CE. The second billion occurred in another 80 years in 1930 CE. The third billion took 31 years to 1961 CE. In another 15 years the fourth billion was past in 1976 CE. By now, world population is at 7.5 billion and expected to pass 11 billion in the next decades. Feeding more and more people is a challenging problem. In the past, science and technology have been able to keep pace by increasing crop protein production, and agricultural yield with new crops and fertilizers. But fertilizers, through runoff, create increased pollution of water. However, solving the problem postpones starvation to the future to be faced again under worse conditions with more population.

Areas with the most advanced food production like the United States, Canada, and Australia are providing half the world's food supply on the export market. However, the major human population concentration is in the tropical areas of Asia, Africa, and South America. Those areas of high population are the most likely to suffer major starvation. Attempts to increase food production are consuming enormous resources, necessitating depletion of energy and materials, which adds to the pollution of air, land, and water. Even the ocean has increasing amounts of trash pollution, as well as, decreasing resources from over fishing. Opposition to scientific progress in food production, like genetic modification, and in new energy generation, threatens to prevent necessary advances. The command in Genesis 1:28 to "be fertile and increase, fill the earth and master it" has been interpreted as license to overpopulate and destroy the planet.

Can science and technology again postpone catastrophe? The entire Earth has been affected with hundreds of species of insects, fishes, amphibians, reptiles, birds, and mammals already extinct in only the recent past. Many more are on the verge of extinction. Five major extinctions have occurred in the history of life on the planet. Another is now in progress as acknowledged by many (Kolbert, 2014). The capacity of the Earth is not infinite and may have already been exceeded (Cohen, 1995).

If species success has any meaning, it must be long term existence. The dinosaurs existed for about 165 million years before going extinct. Humans, so far, have only about a 5 million-year existence. If the history of life were compared to the time scale of one day, human existence would be equivalent to the last minute, with recorded history occupying the last quarter second of the entire day. The term of Human history has not been long in the totality of life. Thus, success has not yet been achieved. Many predictions are dire.

Homer in the Iliad (Pope translation) stated:

> "Like leaves on trees the race of man is found,
>
> Now green in youth, now withering on the ground:
>
> Another race the following spring supplies,
>
> They fall successive and successive rise."

Psalms, 103, 15:16 says: "As for man his days are as grass: as a flower of the field, so he flourishes. The wind passes over it, and it is gone; and the place thereof shall know it no more."

In 1950, Enrico Fermi proposed a paradox. The Fermi Paradox questions, if there is a high probability for the likely existence of intelligent life on the many other planets, why has there been no contact by aliens with humans? Fermi and others who, pursued the paradox reviewed by Webb (2015), conclude that intelligent civilizations are short lived, because they destroy themselves by factors including nuclear war, environmental destruction, resource depletion, over population, climate collapse, and conflict with artificial intelligence.

To conclude: The Biblical King Solomon discoursed about animals (I Kings 5, verse 13). That reference has led to the idea that King Solomon talked to animals. The poet Rudyard Kipling wrote in his poem, "The Butterfly that Stamped:"

> "There was never a Queen like Balkis,
>
> From here to the wide world's end:
>
> But Balkis talked to a butterfly
>
> As you would talk to a friend.

There was never a King like Solomon,
Not since the world began:
But Solomon talked to a butterfly,
As a man would talk to a man.

She was Queen of Sabea—
And he was Asia's Lord—
But they both of 'em talked to butterflies
When they took their walks abroad!"

This volume, lacked the apparent ability of either Queen Balkis (Queen of Sheba) or King Solomon to talk to animals, but tried to do better than talking to animals, by attempting to learn from them.

References

Cohen, J. E. (1995). *How many people can the Earth support?* W. W. Norton & Co., New York, N. Y.

Fromm, E. (1941). *Escape from Freedom.* Farrar & Rinehart, New York.

Happer, W. (2012). "Global Warming Models Are Wrong Again," *Wall Street Journal,* March 27, http://online.wsj.com/articles

Hoffer, E. (1951). *The True Believer: thoughts on the nature of mass movements.* Republished (2010) by Harper. Perennial Modern Classics, New York.

Kennedy, C. (2013). "Why did Earth's surface temperature stop rising in the past decade?" http://www.noaanews.noaa.gov/stories2013

Kipling, R. "The Butterfly that Stamped." https://www.poetryloverspage.com/poets/kipling/just_so_sto ries.html

Kolbert, E. (2014). *The sixth extinction.* Henry Holt and Company, New York, N. Y.

Krantz, G. S. (1970). "Human activities and megafaunal extinctions." *American Scientist,* **58**, 2, 164–170.

Lindzen, R. (2017). "Straight talk about climate change." *Academic Questions,* **30**: 419–432. National Association of Scholars, Springer-Science + Business Media, New York.

NOAA (2017). National Centers for Environmental Information, Climate at a Glance. www.ncdn.noaa.gov/cag/time-series/global/globe/land_ocean/1/6/1880–2017.

Ravolinson, G. (1952). *Herodotus, The History*. In *The Great Books of the Western World*, R. M. Hutchins, ed., Chicago: Encyclopedia Britannica, VI, 1–341 (450 B.C.), Part 2, 1–49.

Webb, S. (2015). *If the Universe is teeming with aliens … Where is everybody?* Springer International Publishing, Switzerland.

Index